# Spatial Variability in Environmental Science - Patterns, Processes, and Analyses

*Edited by John P. Tiefenbacher and Davod Poreh*

Published in London, United Kingdom

IntechOpen

*Supporting open minds since 2005*

Spatial Variability in Environmental Science - Patterns, Processes, and Analyses
http://dx.doi.org/10.5772/intechopen.87655
Edited by John P. Tiefenbacher and Davod Poreh

Contributors
Gunnar Schade, Kristen Koch, Anthony Filippi, Garrison Goessler, Burak Güneralp, Rosa Maria Ceron Breton, Julia Griselda Cerón Bretón, Reyna Del Carmen Lara Severino, Marcela Rangel Marrón, María De La Luz Espinosa Fuentes, Simón Eduardo Carranco Lozada, Lizbeth Cisneros Rosique, Joao Carvalho, Brianna Lunardi, Alex Smith, Chris Houser, Elizabeth George, Arnaud Hequette, Olivier Cohen, Sameh Abdelazim, David Santoro, Mark Arend, Fred Moshary, Sam Ahmed, Marlon Subido Alejos, Alfred Cal, Manuela Magalhães, Selma Pena, Augusto E. Serrano, Jr., Yashier U. Jumah, Rey dela Calzada, Cyril Tom B. Ranara, Jumari C. Fernandez



First published in London, United Kingdom, 2020 by IntechOpen
IntechOpen is the global imprint of INTECHOPEN LIMITED, registered in England and Wales, registration number: 11086078, 5 Princes Gate Court, London, SW7 2QJ, United Kingdom
Printed in Croatia

British Library Cataloguing-in-Publication Data
A catalogue record for this book is available from the British Library

Additional hard and PDF copies can be obtained from orders@intechopen.com

Spatial Variability in Environmental Science - Patterns, Processes, and Analyses
Edited by John P. Tiefenbacher and Davod Poreh
p. cm.
Print ISBN 978-1-83962-459-9
Online ISBN 978-1-83962-460-5
eBook (PDF) ISBN 978-1-83962-461-2

For EU product safety concerns:
IN TECH d.o.o., Prolaz Marije Krucifikse Kozulić 3, 51000 Rijeka, Croatia, info@intechopen.com or visit our website at intechopen.com.

# Meet the editors

Dr. John P. Tiefenbacher (Ph.D., Rutgers, 1992) is a professor of Geography at Texas State University. His research has focused on various aspects of hazards and environmental management. Dr. Tiefenbacher has published on a diverse array of topics that examine perception and behaviors with regards to the application of pesticides, releases of toxic chemicals, environments of the U.S.-Mexico borderlands, wildlife hazards, and the geography of wine. More recently his work pertains to the adaptation to climate change, spatial responses of wine growing to climate change, the geographies of viticulture and wine, artificial intelligence and machine learning to predict patterns of natural processes, historical ethnic enclaves in American cities and regions, and environmental adaptation by European immigrants to North America.

Dr. Davod Poreh Ph.D. is a member of DIETI Department of Electrical Engineering in University of Naples Federico II. He has been working in radar satellite/airborne remote sensing, laser remote sensing, and passive satellite remote sensing in many European countries including Germany and Italy, consequently studying and producing many radar interferograms, persistent scatterers, polarimetiric data sets, airborne/sea laser data, topographic data, etc. His main research interests are in the field of electromagnetic, microwave and passive remote sensing, image processing (optical and SAR), radar, polarimetric SAR, interferometreic SAR, sensor design, antenna, polarimetric based radar simulations, InSAR time series (persistent scatterers interferometry, and SBAS), ground penetrating radar, seismic, gravity, electromagnetic geophysics, information retrieval for land, oceanic, and urban area, EO programming, and GIS.

# Contents

# Preface

This book highlights the importance of spatial perspectives to the discernment of patterns within the physical world and ramifications for those patterns for human activities and for the things that people value. Eight teams of authors have provided examples of their work to demonstrate the relevance of spatial variability to environmental research. Eight chapters are presented in four sections reflecting segments of the environmental sciences – atmospheric science, geomorphological science, biological science, and landscape science. This preface will provide a brief overview of the contents of those chapters.

## The Meteorological Sciences

Our first chapter, "Coherent Doppler Lidar for Wind Sensing" is by Sameh Abdelazim, David Santoro, Mark F. Arend, Fred Moshary, and Sam Ahmed. In it, the authors detail the design, operation, and testing of a new technology, coherent Doppler lidar to monitor wind. The need to assess wind direction, speed, and trajectory is vital to safely interacting with the atmosphere in many ways. The application of this new device could be used to detect and track dangerous changes in wind speeds and wind directions. It may improve safety for flying and structural engineering. It may also enable earlier warning to communities regarding the development of dangerous tornadic and cyclonic storm conditions. They demonstrate an approach to discerning the spatial variability of wind.

The atmosphere can generate other hazardous conditions like the problem of differential urban heating, which may yield dangerous thermal conditions for vulnerable populations. The second chapter, "Low-key Stationary and Mobile Tools for Probing the Atmospheric UHI Effect" by Kristen Koch, Gunnar W. Schade, Anthony M. Filippi, Garrison Goessler, and Burak Güneralp, describes a study that the authors designed to track and monitor urban heat islands, a phenomenon related to modification of natural light-reflecting and scattering land covers to enhance thermal absorbance and reradiation of heat, urban activities that generate particulates and other heat-absorbing pollutants, and conversion of land uses to extend the size of the urban footprint. The authors detail the design and operation of a spatially transferable research activity that can be adapted to meet the dual value of educational instruction (of undergraduate and graduate students) and meaningful data acquisition and analysis. Their research involves field work, remote sensing analysis, and data analysis of local and regional atmospheric monitoring records. They demonstrate an approach to discerning the spatial variability of thermal energy in the boundary layer of the atmosphere.

In the third chapter, Rosa María Cerón Bretón, Julia Griselda Cerón Bretón, Reyna del Carmen Lara Severino, Marcela Rangel Marrón, María de la Luz Espinosa Fuentes, Simón Eduardo Carranco Lozada, and Lizbeth Cisneros Rosique present a study of "Mapping and Estimation of Nitrogen and Sulfur Atmospheric Deposition Fluxes in Central Region of the Mexican Bajio." Monitoring the ramifications of human activities, particularly burning fossil fuels and other industrial activities, is important not only for human health and welfare, but also for the security and safety of the ecological systems upon which we depend. Sulfur and nitrogen are two components of the atmospheric pollution regime that can have both human and ecological effects that are often determined by the molecular forms they take, but are also a function of concentrations, timing, and spatial distributions that effect exposure to harmful (if not toxic) levels of contamination. The authors study the phenomenon of deposition of these elements in the city of León, Mexico. They describe the methods used to compile and analyze data from regional monitoring networks and samplers. The implications of these contaminants for local and regional flora and ecosystem health is discussed. They demonstrate an approach to discerning the spatial variability of pollution in the lower atmosphere.

## Geological Sciences

Change in the environment poses certain hazards in certain places. The capacity to quickly and accurate assess the results of an extreme natural event is an ever-present desire. The coastal zone is one of the most dynamic environments on Earth as it is in constant flux due to the interactions between the land, the atmosphere, and the ocean. Evaluation of the impacts of storms on barrier islands presents certain challenges, but our fourth chapter, "Monitoring Storm Impacts on Sandy Coastlines with UAVs" by Alex Smith, Brianna Lunardi, Elizabeth George, and Chris Houser presents an effort to use unmanned aerial vehicles to do just that, with speed and precision as its goal. The authors describe the methods, challenges, and limitations of their new approach in the context of Prince Edward Island, Canada. Their study reveals the effectiveness of linear, aerial, and volumetric monitoring of the spatial changes of dunes after a storm using drone technology. They demonstrate an approach to discerning the spatial variability of coastal geomorphological processes.

Similarly focused on the spatial variability of coastal geomorphology, our fifth chapter "Recent Advances in Coastal Survey Techniques: From GNSS to LiDAR and Digital Photogrammetry: Examples on the Northern Coast of France" examines the effectiveness of the technologies used to gather spatial data. Olivier Cohen and Arnaud Héquette discuss the benefits and limitations of several sensor technologies to gather land surface measurements from which digital terrain models (DTMs) can be derived. The authors employ these methods in case studies along France's northern coast. They demonstrate the implications of the selection of technologies on analysis of the spatial variability of coastal geomorphology.

## Biological Sciences

The sixth chapter of this volume is by João Carvalho, Manuela Magalhães, and Selma Pena and covers the analysis of forest ecology for landscape planning. In "Spatial and Temporal Variability Regarding Forest – From Tree to the Landscape," the authors discuss the ramifications of scale on the assessment of the integrity, biodiversity, and viability of forests in Portugal. An important lesson described in this chapter is that spatial variability varies with spatial scale. Or to put it differently, one could possibly not see the forest for the trees and that attention to scale is vitally important to assessment. Such mindfulness to spatial scale should guide the selection of the metrics (biodiversity, forest fire risk, disturbance, fragmentation, etc.) by which the conditions of a forest are judged. Interventions in using forested lands must focus on clearly stated management goals and those goals need to vary with spatial scale and by location. They demonstrate approaches to assessments of the spatial variable of the health and sustainability of forests.

In chapter 7, "Ecological and Social Impacts of Aquacultural Introduction to Philippines Waters of Pacific Whiteleg Shrimp Penaeus vannamei" by Marlon S. Alejos, Augusto E. Serrano Jr., Yashier U. Jumah, Rey dela Calzada, Cyril Tom B. Ranara, and Jumari C. Fernandez, we learn of the implications of introduced species for a region's ecosystems. The problem of introduced (i.e., non-native) species for indigenous ecologies and their human communities is a well-known matter. In this study, however, the authors review the evidence of ramifications from aquacultural farming of a species of shrimp from the other side of the ocean, a process of introduction that began more than 50 years ago. They examine the consequences of accidental releases of shrimp into the region's environments: the likelihood of survival and reproduction of escapees, the impacts to native shrimp, the impacts on other biota, the consequences of ecosystems, and human communities in the region. Their analysis of the evidence showed that much remains unknown about the matter, but that "no evidence" does not mean "no impact." They demonstrate the importance of the spatial variability of economic activities tied to disturbance of native species and ecosystems.

## Landscape Sciences

The eighth and final chapter takes us into the built environment and the concern with spatial variation in the landscape. Mapping of the engineered, or "built" landscape, is another important component of the ability to map and understand the spatial patterns of the human environment and their implications for humans' interactions with nature. "High-resolution Object-based Building Extraction Using PCA of LiDAR nDSM and Aerial Photos" by Alfred Cal presents a methodology by which LiDAR nDSM and aerial photos can be analyzed with principal components analysis (PCA) to generate a highly successful and highly accurate extraction of building footprints. The author demonstrates the manipulations that are needed to map human structures in a regional rural to semi-rural landscape in Belize. He demonstrates an approach that improves our capacity to assess the spatial variability of land cover and land uses in our natural environments.

It is our hope that these chapters provide interesting reading for students who are compelled by questions of spatial variability (or Geography) in the realms of diverse environments of our planet.

**John P. Tiefenbacher**
Texas State University,
USA

**Davod Poreh**
University of Naples Federico II,
Italy

Section 1

# Meteorological Sciences

Chapter 1

# Coherent Doppler Lidar for Wind Sensing

*Sameh Abdelazim, David Santoro, Mark F. Arend, Fred Moshary and Sam Ahmed*

## Abstract

An eye-safe all-fiber Coherent Doppler Lidar for wind sensing system has been developed and tested at the Remote Sensing Laboratory of the City College of New York, New York, NY. The system, which operates at a 20 kHz pulse repetition rate and acquires lidar return signals at 400 MSample/s, accumulates signals that are as much as 20 dB lower than the receiver noise power by using embedded programming techniques. Two FPGA embedded programming algorithms are designed and compared. In the first algorithm, power spectra of return signals are calculated and accumulated for different range gates. Line of sight wind speed estimates can then be calculated after transferring the range gated accumulated power spectra to a host computer. In the second FPGA algorithm, a digital IQ demodulator and down sampler allow an autocorrelation matrix representing a pre-selected number of lags to be accumulated. Precision in the velocity measurements is estimated to be on the order of 0.08 m/s and the precision in the measured horizontal wind direction is estimated to be to be about 2°.

**Keywords:** Doppler Lidar, wind sensing, heterodyne detection, FPGA, coherent laser, eye-safe

## 1. Introduction

The first wind measurement device (anemometer) was invented in 1450 by an Italian architect named Leon Battista Alberti. Four hemispherical cups anemometer was later invented in 1846 by Dr. John Thomas Romney Robinson. Today, wind speed and direction can be measured by using classical anemometers, sonic anemometers, rawinsondes, SODAR (Sonic Detection and Ranging), RADAR (Radio Detection and Ranging), and LIDAR (Light Detection and Ranging). Sonic anemometers determines instantaneous wind speed and direction by measuring how much sound waves traveling between a pair of transducers are sped up or slowed down by the effect of wind. SODAR measures wind speed through measurements of the scattering of sound waves by atmospheric turbulence. Both radar and LIDAR use similar technique such as SODAR but instead of using sound waves, radar uses microwave, and LIDAR uses laser waves.

Sodars and radars are used in wind profilers to measure wind speed and direction at various altitudes above ground level. Wind speed can be estimated by transmitting five beams; one is vertical to measure vertical wind velocity, and the other beams are orthogonal to each other to measure horizontal components of the wind. The profiler's assumption to measure wind speed is that turbulent eddies that scatter probing signals are carried along by the mean wind. **Figure 1** shows radar and sodar wind profilers that are mounted at Liberty Science Center, New Jersey, and on top of the Metlife building in the center of Manhattan, New York, respectively, as part of the New York City meteorological network (NYC MetNet). These types of instruments are large and not portable.

Coherent Doppler Lidar (CDL) has proven to be a powerful tool for remote sensing of the atmosphere, and has been widely adopted in applications such as measuring atmospheric wind velocity, turbulence, aerosol concentration, cloud height and velocity, and detection of atmospheric constituents and pollutants. The CDL systems have been developed for remote sensing measurements since the late 1960's. The first CDL wind-sensing system was reported by Huffaker et al. [1], where a 10.6 μm cw $CO_2$ laser was used. $CO_2$-laser-based CDL systems have been used for airborne clear air wind and hard target measurements such as ranging and produced valuable results for a long time. Since the late 1980s, CDL systems with newly developed solid-state lasers attracted a lot of researchers due to advantages of size, weight, reliability, and lifetime [2].

Operation at shorter wavelengths allows for higher spectral resolution, which means higher velocity resolution. A lot of effort has been put into 2 μm pulsed systems mainly intended for wind measurements [3–5]. Kavaya et al. [6] developed a 1.06 μm pulsed CDL system that realized a measurable range of a few tens of kilometers used for launch-site wind sensing. Karlsson et al. [3] reported a 1.5 μm cw all-fiber wind sensing CDL system, which utilized optical fiber components used in telecommunication systems.

In this study, we report on the design, measurements, and performance of a CDL system for wind measurements [7]. The design involves a very low energy per pulse (12 μJ/pulse), because it employs all-fiber optic laser components for availability, cost affordability, robustness, and size compactness. As a result of this low energy per pulse, a very high frequency repetition rate (FRP) is used, which produces a very large volume of returned signals. Acquiring such a large volume of data at a very high sampling rate and processing it cannot be achieved using a classical data acquisition and processing hardware. Therefore, signal pre-processing has to be carried out on hardware level by means of the FPGA, which allows for real time processing and a moderate data transfer rate from the data acquisition card to the host PC. In our design, signal pre-processing was implemented to produce the power spectrum of time gated received signals by calculating fast Fourier transform FFT, which produces fixed spatial resolution gates. Another pre-processing technique was implemented that involved the calculation of received signals' autocorrelation, which can be used to find the power spectrum at any desired range resolution.

The system consists of a distributed-feedback (DFB) semiconductor laser emitting at a 1.5 μm in addition to an erbium-doped fiber amplifier (EDFA) in a master

**Figure 1.**
*A radar wind profiler (left) mounted on the liberty science center and a sodar wind profiler (right) mounted on a NYC high rise [8].*

oscillator power amplifier (MOPA) configuration. A notable feature of our system is that it utilizes polarized maintained (PM) fiber optics, which ensures the maintaining of polarization state between both local oscillator and back scattered fields. The advantage of using a 1.5 μm laser source is the eye-safety feature, which allows for operation in urban areas. In addition to eye-safety feature, the usage of a 1.5 μm source allows the system to benefit from the technology and component development driven by the telecommunication industry, which results in significant cost deductions.

In this study, the development and operation of a CDL system for wind sensing is presented. In Section 2, the system configuration is presented and system's main components are described. In Section 3, transceiver noise analysis and coherent lidar signal range dependence are examined. In Section 4, signal processing and FPGA programming is introduced. In Section 5, wind measurement results are reported in both vertical pointing and scan modes.

### 1.1 Coherent Doppler lidar theory

In coherent Doppler lidar, laser pulses are transmitted into the atmosphere and interact with aerosols within the atmosphere. As a result of this interaction, laser signals are backscattered towards the laser source, which can be detected and measured through an optical detector. According to the movement of the atmospheric aerosols with respect to the laser source, backscattered signals may suffer a frequency shift (Doppler shift) that is proportional to velocity of moving aerosols. The Doppler shift $\Delta f$ of laser signals with $\lambda$ wave length is given by:

$$\Delta f = \frac{2\nu}{\lambda} \tag{1}$$

where $\nu$ is the velocity of the aerosols, i.e. wind velocity.

### 1.2 Heterodyne detection theory

In heterodyne optical detection, local oscillator and backscattered signals are optically mixed through an optical coupler. The resulting mixed signal is then incident upon a photodetector. Both local oscillator and backscattered fields can be represented as:

$$x_{lo} = A_{lo}\, cos\,(\omega_o t) \tag{2}$$

$$x_s = A_s\, cos\,(\omega_o t + \omega_s t) \tag{3}$$

where $\omega_0$ is the local oscillator frequency and $\omega_s$ is the frequency shift that backscattered signals may suffer. The optical intensity as seen by the heterodyne detector is given by:

$$\begin{aligned} I_{opt} &= [A_{lo}\cos(\omega_o t) + A_s \cos(\omega_o t + \omega_s t)]^2 \\ &= \frac{{A_{lo}}^2}{2} + \frac{{A_s}^2}{2} + A_{LO}A_s\cos(\omega_s t) + (\text{High frequency component}) \end{aligned} \tag{4}$$

The terms: $2\omega_0$, $2(\omega_0 + \omega_s)$, and $(2\omega_0 + \omega_s)$ are at higher frequencies than detector's bandwidth and will not be seen by the detector. As a result, the generated photodiode electric current will equal to:

$$I_d = \eta A I_{opt} \tag{5}$$

$$= \frac{\eta A {A_{lo}}^2}{2} + \frac{\eta A {A_s}^2}{2} + \eta A A_{lo} A_s \cos(\omega_s t) \tag{6}$$

$$= \eta P_{lo} + \eta P_s + 2\eta\sqrt{P_{lo} P_s}\cos(\omega_s t) \tag{7}$$

where: $A$ and $\eta$ are detector's surface area and photo responsivity, respectively. $I_d$ consists of a dc component = $\eta P_{lo}$ + $\eta P_s$ and an ac component = $2\eta\sqrt{P_{lo}P_s}\cos(\omega_s t)$. since $P_{lo} >> P_s$, then:

$$I_{d(dc)} = \eta P_{lo} \tag{8}$$

$$I_{d(ac)} = 2\eta\sqrt{P_{lo}P_s}\cos(\omega_s t) \tag{9}$$

Signal power can be calculated as:

$$< i_s^2 > = \left(I_{d(rms)}\right)^2 \tag{10}$$

$$= 2\eta^2 P_{lo} P_s \tag{11}$$

Detector's responsivity is related to detector's quantum efficiency through the following relationship:

$$\eta = \frac{e\eta_q}{h\upsilon} \tag{12}$$

where; $e$ is electron charge, $\eta_q$ is the quantum efficiency of the detector, $h$ is Plank's constant, $\nu$ is laser's frequency,

$$< i_s^2 > = 2\left(\frac{e\eta_q}{h\upsilon}\right)^2 P_{lo} P_s \tag{13}$$

In this system, backscattered signals are sampled at 400 MHz using a 14-bit ADC equipped with an on-board FPGA. The laser pulse frequency rate (PFR) is 20 kHz, which limits the maximum measurement range to 7.5 km. To estimate wind velocity, the frequency shift of scattered signals (Doppler shift) has to be extracted. Backscattered signals are broken into time gates to represent desired range distances and a power spectrum of each range gate is calculated by Fast Fourier Transform (FFT). A gate length of 128 data samples is chosen, which corresponds to 48 m range distance. Due to low pulse energy (14 μJ/pulse), power spectrum accumulation is needed to improve detection probability and velocity estimation accuracy. The wind velocity of each range gate is estimated from the calculated mean frequency of a post processed power spectrum around the peak frequency.

## 2. SYSTEM'S configurations

In this section, the main system components are presented and an over view of the system's operation is explained. System's configurations and testing of optical and electrical components are also presented in this section.

### 2.1 System overview

The system configuration is shown in **Figure 2**, which consists of the following components: (i) Laser source (ii) Accousto-optic modulator (iii) Fiber amplifier (iv) Optical circulator (v) Optical antenna (vi) Balanced detector, and (vii) Signal

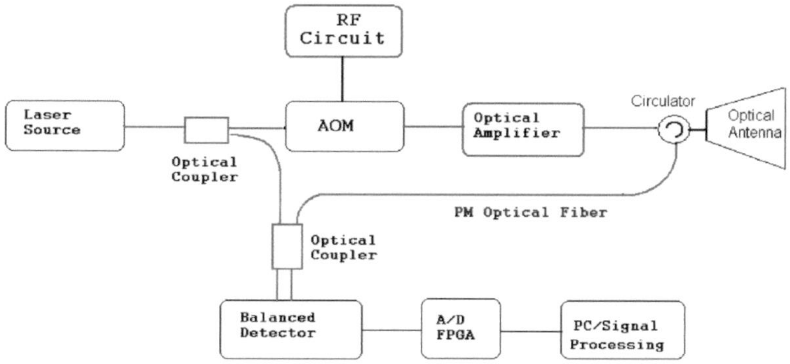


**Figure 2.**
*Coherent Doppler Lidar system's configuration.*

processor. Optical components are connected with a single mode polarized maintained (PM) optical fiber.

Our laser source has two outputs: a low power seed laser that is used as a local oscillator (LO), and a high power output (0.5 W) that is modulated, pulsed, and frequency shifted using an acousto-optic modulator (AOM). Electronic circuits drive the AOM to shift laser signals by 84 MHz and generate 200 ns Gaussian shaped laser pulses. These laser pulses are amplified through an erbium doped fiber amplifier (EDFA) then transmitted from port 1 to port 2 of the optical circulator. To minimize the back reflection from port 2 back to port 1, the fiber tip at port 2 is angled and polished. Laser pulses are transmitted into the atmosphere and aerosol particles scatter the laser signals back into the lens, which in turn are transmitted from the optical circulator's port 2 to port 3. Backscattered and LO signals are optically mixed using an optical coupler. Optically mixed signals are heterodyne detected through an optical balanced detector, which generates RF signals. These RF signals are acquired at a 400 MHz sampling rate using an analog to digital converter card (ADC), which is equipped with an on-board field programmable gate array (FPGA) to allow for real time analysis. Digital data is then streamed to a host PC for further processing. More detailed explanations of key components are presented below.

### 2.2 Laser source

The laser source is a distributed feedback erbium doped fiber laser (DFB-EDFL) from NP Photonics. The laser's wavelength is 1545.2 nm, and it has two outputs; first output, used as a seed laser, and second output has an adjustable output power up to 500 mW. The spectrum of the delayed heterodyne detected signal as measured by a spectrum analyzer has a full width at half-maximum (FWHM) of a few kHz. The laser linewidth is approximately 3 kHz, which corresponds to a velocity estimation accuracy of 0.2 cm $s^{-1}$, so the laser linewidth is enough for our specification.

### 2.3 AOM

The continuous wave (CW) laser input is frequency shifted and pulsed through the AOMs, where an ultrasonic pulse is generated at a piezoelectric device by driving RF signals. Two AOMs are connected in series to obtain a very low extinction ratio. Each AOM shifts the frequency by 42 MHz, which leads to a total frequency shift of 84 MHz. The purpose of shifting the frequency of transmitted signals by 84 MHz is to shift the frequency of the zero velocity, so that both positive and negative Doppler shifts could be recognized. The driving RF signals, turn the

AOMs on during the 300 ns of the 20 kHz RF driving pulse to generate a laser pulse with 200 ns Full Width at Half Maximum (FWHM).

### 2.4 Fiber amplifier

An erbium-doped fiber amplifier (EDFA) that has an average power of 340 mW and a peak power of 74.6 W is used. This peak value cannot be increased beyond 74.6 W because of the Stimulated Brillouin Scattering (SBS), which can take place when an intense laser beam travels through a medium such as an optical fiber. SBS is generated from the acoustic vibrations in the medium that are caused by variations of the electric field of a traveling laser beam. Usually a laser beam undergoes SBS in an opposite direction to the incoming beam, which in our case can go back to the laser amplifier and cause damage to it. The EDFA has two amplifier stages; pre-amplifier and power amplifier. Output power is adjusted in a current control mode by adjusting the current of the power amplifier stage.

### 2.5 Optical circulator

The optical circulator ensures that the amplified output laser pulse is transmitted into the optical antenna and not into the detector. It also ensures that received backscattered signals and signals reflected off the fiber tip from the output pulse are directed into the receiver and not into the fiber amplifier. The back-reflection signal level at the fiber tip of the output port (port 2) of the optical circulator is very critical, because it can damage the optical detector.

### 2.6 Optical antenna

A 4″ diameter lens with a focal length of 50 cm is used. The truncation ratio is approximately 0.88, and the lens' Rayleigh range is approximately 5 km, which means that the laser beam can be collimated for all desired range (100 m to 4 km). This lens is mounted on an aluminum rail with a fiber holder that houses the optical fiber. A 6″ mirror is also mounted on the same rail to steer the laser beam, and the entire setup is mounted on optical table.

### 2.7 Balanced detector

An InGaAs heterodyne balanced detector with a bandwidth extending from d.c. to 125 MHz is used to retrieve backscattered signals. The benefit of using a balanced detector is to subtract the two optical input signals from each other, which results in the cancelation of common mode noise. This allows for detection of small changes in the signal path from the interfering noise floor. The photo current is converted into voltage through the detector's transimpedance amplifier module. Detector's noise was measured using a spectrum analyzer while no optical signals were applied to its inputs, gain was set to 1×, transimpedance was set to 1.4 kΩ, coupling was set to DC, and spectrum analyzer's frequency resolution was set to 3 MHz. Detector's noise was equal to −83 dBm, which is 1 dB below detector's specification of 3.6 pW $Hz^{-1/2}$. The detector has a non-flat gain response, **Figure 3**, i.e. the gain of the detector varies with the frequency of the input signals.

To correct for this non-flat gain shape, received signals' power spectrum is divided by the power spectrum of detector's output while no signal is present. The measured power when signal is present can be represented as:

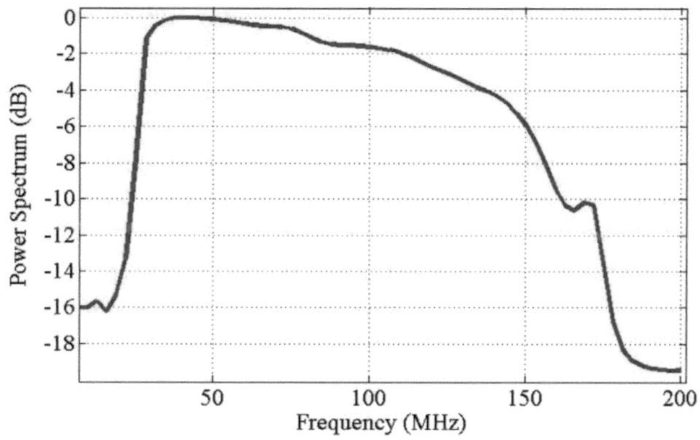


**Figure 3.**
*Non-flat gain response of the heterodyne balanced detector.*

$$P_{total} = P_{sig} + P_{noise} \tag{14}$$

Dividing this measured power spectrum by the power spectrum where no signal is present gives:

$$\frac{P_{total}}{P_{noise}} = 1 + \frac{P_{sig}}{P_{noise}} \tag{15}$$

$$= 1 + SNR \tag{16}$$

where SNR: is the signal to noise ratio.

Subtracting 1 from Eq. (16), gives the SNR. The previous technique is used in our signal processing to estimate backscattered signal power.

### 2.8 Polarized maintained (PM) fiber optic

All optical components are connected through PM optical fibers to ensure that the polarization state of the electric field of the local oscillator and that of the backscattered signals are very close, if not the same. Maintaining the polarization state throughout the different components of the system ensures a high level of the heterodyne detected signals.

## 3. Power analysis and SNR range dependence

In this section, noise components of the heterodyne photodetector are analyzed, detailed SNR analysis is presented, and the optimum local oscillator power level is determined. The range dependence of SNR is also investigated, and system performance is evaluated. Analytical and experimental wideband SNR was compared. In Section 1, we present the SNR at the heterodyne detection and in Section 2 we present the range dependence of the wideband SNR.

### 3.1 Transceiver noise analysis

In optical heterodyne detection, The SNR of the Lidar system determines the system's ability to detect low level backscattered signals out of noise [9, 10]. Lidar heterodyne photoreceiver optimization is required to increase the receiving sensitivity [11].

It was shown that heterodyne detection sensitivity can reach its maximum value if local oscillator power is set to an optimum level [10]. The following analysis presents different heterodyne photodetector's noise components and gives an estimate to SNR.

The noise at the output of the optical detector consists of: (1) thermal noise (Johnson noise), (2) shot noise due to local oscillator induced current, and (3) laser's relative intensity noise (RIN).

Thermal noise is related to the detector and does not depend on the local oscillator power ($P_{lo}$). Thermal noise is expressed as:

$$< i_{th}^2 > = \frac{4kTB}{R_l} \tag{17}$$

where: $k$ is Boltzmann's constant, $T$ is temperature in degrees Kelvin, B is detector's bandwidth, and $R_l$ is detector's load resistor.

Shot noise, unlike signal powers that cancel through the balanced detector, the uncorrelated shot noise adds [12], resulting a mean-square noise at the output of the detector given by:

$$< i_{sh}^2 > = 2eiB \tag{18}$$

where: $i$ is the detector's current caused by the local oscillator power. This current can be calculated as follows:

$$i = en_e \tag{19}$$

where, $n_e$ is the number of electrons, which is given by:

$$n_e = \eta_q e_{ph} \tag{20}$$

where, $n_{ph}$ is the number of photons incident on the detector, $\eta_q$ is detector's optical efficiency.

$$e_{ph} = \frac{P_{lo}}{h\nu} \tag{21}$$

$$\therefore < i_{sh}^2 > = \frac{2\eta_q e^2 BP_{lo}}{h\nu} \tag{22}$$

Laser relative intensity noise (RIN) is a property of the laser source, which is related to square value of local oscillator power through the following relationship:

$$< i_{RIN}^2 > = (R_{in})R_b \left(\frac{e\eta_q}{h\upsilon}\right)^2 B(P_{lo})^2 \tag{23}$$

where: $R_b$ is *RIN* suppression ratio through the use of balanced detection. The SNR can now be expressed as:

$$SNR = \frac{< i_s^2 >}{< i_{th}^2 > + < i_{sh}^2 > + < i_{RIN}^2 >} \tag{24}$$

$$= C\left[1 + \frac{2kTh\upsilon}{\eta_q e^2 P_{lo} R_l} + \frac{\eta_q R_{in} R_b P_{lo}}{2h\upsilon}\right]^{-1} \tag{25}$$

where: $C$ is an independent term of local oscillator power = $\frac{\eta_q}{Bh\upsilon}P_s$. When $P_{lo}$ is small, the second term in the denominator of Eqs. (3)–(9) dominates, and *SNR* is directly proportional to $P_{lo}$. Hence SNR increases with increasing $P_{lo}$. On the other hand, when $P_{lo}$ is large, the third term in the denominator dominates, and *SNR* is inversely proportional to $P_{lo}$. The *SNR* then decreases with increasing $P_{lo}$. This means that *SNR* will increase as local oscillator power increases until it reaches a maximum value (where $P_{lo}$ is optimum), after which it starts to decrease. The optimum value of $P_{lo}$ can be determined by plotting the SNR as a function of $P_{lo}$ assuming room temperature, $R_{in} = -152$ dB as provided by our laser vendor, $R_b = -25$ dB, $\eta_q = 0.8$, and $R_l = 50\ \Omega$. It is shown that *SNR* is maximum when $P_{lo}$ is approximately = 10 mW, however, we chose to set $P_{lo}$ to approximately 5 mW to avoid operating the detector near its damage threshold, **Figure 4**.

### 3.2 Coherent Lidar signal range dependence

In this section, we study the range dependence of SNR for the coherent laser radar (CLR) heterodyne detection using a mono-static configuration, and we compare analytical and experimental results. Mono-static configuration was believed to have an improved performance due to the correlation of the transmitted and back scattered fields. This correlation is the result of wave-front tilts self correction in a mono-static configuration [13–15]. The SNR range dependence of a CLR mono-static system is evaluated by using the concept of backprojected local oscillator (BPLO), which is the imaginary local oscillator field distribution projected at the target side of the receiver aperture, receiver lens, originating from the detector [13, 16, 17]. Frehlich and Kavaya [13] derived an equation that describes SNR as a function of range assuming a Gaussian Lidar system i.e., transmitter and LO fields are deterministic, detector response function is uniform, and the detector collects all LO and backscattered power incident on the receiver aperture. The SNR was then found by calculating the overlap integral between the BPLO and the backscattered fields on the receiver plane assuming a distributed aerosol target assuming ideal conditions, i.e. shot noise limited detector and a deterministic beam. To take into account the effects of refractive turbulence on CLR performance, different techniques of wave propagation in random medium were used [18]. Analysis shows that the SNR is proportional to the product of direct detection power and heterodyne efficiency. The calculation of received power and SNR requires mutual coherence function of the backscattered field incident on the receiver. As for natural aerosol targets, backscattered field at each aerosol particle has a random phase, and the mutual coherence function of the total backscattered

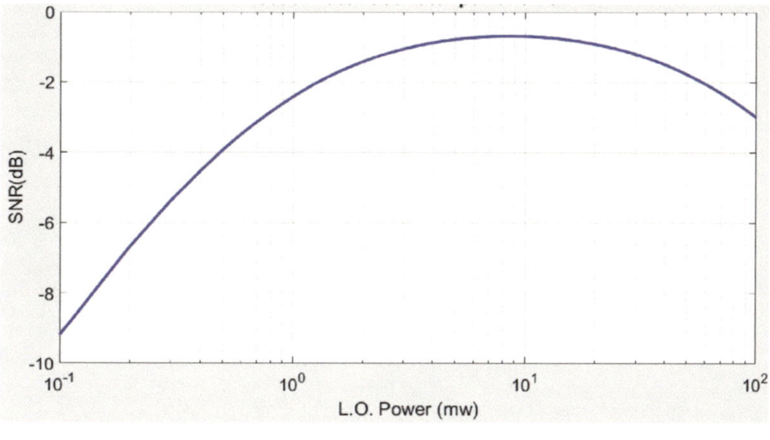


**Figure 4.**
*Normalized* SNR *as a function of local oscillator power* $P_{lo}$.

field is the integration of all mutual coherence functions from each aerosol particle. The SNR range dependence equation is expressed as [19]:

$$SNR(L) = \frac{\eta_D(L)\lambda E\beta K^{2L/1000}\pi D^2}{8hBL^2} \tag{26}$$

where; $\eta_D$ is the system efficiency, given by:

$$\eta_D(L) = \frac{\eta_{total}}{\left\{1 + \left(1 - \frac{L}{L_F}\right)^2 \left(\frac{\pi (A_C D)^2}{4\lambda L}\right)^2 + \left(\frac{A_C D}{2S_O(L)}\right)^2\right\}} \tag{27}$$

where; the parameters of Eqs. (26) and (27) are introduced in **Table 1**.

The performance of a 4″ diameter antenna was evaluated both theoretically and experimentally while focusing the laser beam at approximately 1.8 km. Continuous 38 range gates having a length of 0.32 μs (48 m range resolution) were obtained between a minimum range of 128 m and a maximum range of approximately 2 km. The power spectra of received signals from 10,000 laser shots were accumulated and wideband SNR was estimated. **Figure 5** shows theoretical and experimental

| Parameter | Descriptions | Value |
|---|---|---|
| L | Range (m) | |
| B | Bandwidth | 100 MHz |
| λ | Wave length | 1545.2 μm |
| E | Pulse energy | 7 μJ |
| D | Effective aperture diameter | 0.15 m |
| τ | Pulse width | 200 ns |
| β | Atmospheric backscatter coefficient | $8.3 \times 10^{-7}$ m/sr |
| K | One way atmospheric transmittance | 0.95 km |
| $L_F$ | Focal range of optical antenna | 1.8 km |
| $A_c$ | Correction factor | 0.76 |
| $Cn^2$ | Refractive index structure constant | $2 \times 10^{-14}$ $m^{-2/3}$ |
| $\eta_{total}$ | Total system efficiency | −2.2 dB |
| $S_o(L)$ | Transverse coherent length | $\sim(1.1\, kw^2 L\, Cn^2)^{-3/5}$ |
| $kw$ | Wave number = 2π/λ | |

**Table 1.**
*Parameters corresponding to analytical estimation of wideband SNR range dependence.*

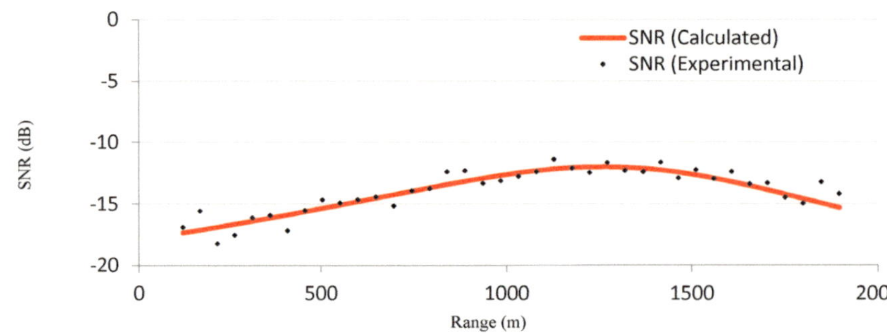


**Figure 5.**
*Wideband SNR range dependence (points, experimental; solid curve, theoretical).*

wideband SNR range dependence. It is clear that both measured and theoretically calculated wideband SNR have a very good agreement. The parameters used in this analysis are listed in **Table 1**.

## 4. FPGA programming and wind measurements analyzed using FFT

For a 20 kHz PFR and a 14-bit ADC with a sampling rate of 400 MHz, data transfer rate from the data acquisition card to the host PC will be 800 Mbyte/s. This high data transfer rate is difficult to be achieved and requires additional hardware and software. Moreover, the amount of data collected in 1 day will be more than 69 Tbyte, which makes data archiving for just a few days nearly impossible. Due to the fast PFR, signal processing on the host computer cannot be achieved in real time, and will cause data to be lost. Therefore, programming the FPGA to calculate power spectra or correlograms of backscattered signals and accumulate the results over a large number of pulses (we chose 10 K pulses) will not only take the burden off the host PC and allow for real time analysis, but will significantly reduce data transfer rate across the PCI express bus to the host PC. In this approach, a signal processing algorithm is implemented and programmed onto the FPGA so that backscattered signals time gating, power spectrum calculation, and accumulation will all be simultaneously carried out on the hardware level as soon as signals are acquired by the ADC. Power spectrum of backscattered signals can be estimated directly by calculating the FFT of the time gated signals, or by calculating the FFT of signals' autocorrelation. FFT pre-processing algorithm and wind measurement results using FFT technique will be explained in the following sections, while autocorrelation pre-processing algorithm and wind measurement results using autocorrelation technique will be explained in details in Section 5.

### 4.1 ADC card

Backscattered signals are sampled at 400 MSPS using a 14-bit ADC card, which features two 14-bit, 400 MSPS A/D and two 16-bit, 500 MSPS DAC channels with a Virtex5 FPGA computing core and a PCI Express host interface. The Virtex5 FPGA can be programmed using VHDL and MATLAB using the Frame Work Logic toolset. The MATLAB Board Support Package (BSP) allows for real-time hardware-in-the-loop development using graphical, block diagram Simulink environment with Xilinx System Generator toolset. Software tools for host PC development can be performed using C++.

### 4.2 FPGA programming algorithms

A signal pre-processing algorithm is initially implemented as a logic design, which can be simulated and tested using Matlab/Simulink software. This logic design is then compiled using Xilinx system generator toolset to produce a hardware VLSI image, which can be downloaded into the FPGA. We chose to pre-process backscattered signals in two different techniques: (a) calculate the FFT of time gated signals then accumulate the resulting power spectrum, and (b) calculate the autocorrelation of the backscattered signals then accumulate the resulting autocorrelation matrix for 10 k laser shots.

### 4.3 FFT pre-processing algorithm

In this pre-processing algorithm, received signals are time gated into portions corresponding to spatial range gates, FFT is estimated for each range gate, and the corresponding power spectra are accumulated over 10 k laser shots. Power spectrum can be calculated using the FFT as follows:

The normalized Fourier transform of a time domain signals *f(t)* can be expressed as:

$$F_T(w) = \frac{1}{\sqrt{T}} \int_0^T f(t)\, e^{-iwt}\, dt \tag{28}$$

The discrete spectral density can then be found as:

$$PSD(w) = \lim_{T \to \infty} E\left[\left|F_T(w)^2\right|\right] \tag{29}$$

Eq. (29) shows that the squared modulus of the Fourier transform is the power spectrum. Therefore, we program the FPGA to calculate the square modulus of the output of the FFT block. Our ADC vendor provided us with an FPGA logic design that streams digitally converted signals (sampled at 400 MHz rate) across the PCI express bus to the host PC. This logic design accepts an external trigger signal to start data acquisition. A 20 kHz signal synchronized with laser pulses is used to trigger the data acquisition process. The ADC card operates in a frame mode in which it acquires a frame of incoming data every time it receives an external trigger's interrupt. A frame size of 8192 samples is chosen, which corresponds to approximately 3.1 km. Xilinx Fast Fourier Transform 7.1 circuit block is used in a pipelined-streaming-io mode to calculate FFT for a vector of 128 samples of time gated scattered signals (corresponding to a 48 m spatial resolution). Logic circuits that calculate the modules of the FFT complex output are also implemented and integrated with this design.

### 4.4 Host computer signal processing

Once accumulated power spectra are streamed from the FPGA across the PCI express bus, data post processing is carried out on the host PC to estimate various parameters such as radial wind velocity, backscattered signal strength, and velocity statistics. Data archiving and visualization are also carried out on the host PC.

The Doppler lidar estimate of the radial component $v$ ($m.s^{-1}$) of the velocity vector is obtained from the mean-frequency $\Delta f$ (Hz) of the Doppler lidar signal as:

$$v = \frac{\lambda}{2} \Delta f \tag{30}$$

where λ (m) is the laser wavelength. As a result, the maximum radial velocity that can be measured is given by:

$$v_{max} = \frac{\lambda}{2} f_{\text{max}}, \tag{31}$$

which is approximately 30 m $s^{-1}$.

The main parameter of interest in Doppler wind measurement is the mean frequency shift of the backscattered signal, because it is directly proportional to the mean velocity of moving aerosol particles within the atmosphere [20, 21]. Doppler frequency shift can be estimated by finding the centroid of the discrete power spectrum of the backscattered signal after removing the amplifier gain shape [22]. One easily calculated method of finding this frequency from a discrete power spectrum is to find the frequency of the highest power, i.e. the frequency corresponding to the peak power [23]. If the backscattered signal's mean frequency shift and the frequency corresponding to the peak power do not coincide, the velocity estimate can be off by as much as one-half of a frequency resolution.

### 4.5 Setup procedure to observe scattered signals

Initially the following procedure was followed when measurements were taken at the remote sensing laboratory of the CCNY to assure the fiber's optimum alignment with respect to the lens. In this procedure, the laser pulses were shot at a hard target (~100 m), and the scattered signal was obtained and monitored on the oscilloscope. **Figure 6** shows the scattered signal off a hard target in time domain. Maximizing the magnitude of the hard target's scattered signal through adjusting fiber's x, y, and z positions achieves the optimum fiber's location to focus the beam at about 100 m (the hard target's distance). To focus the beam at a different distance, the z position of the fiber holder can be adjusted accordingly. We then direct the laser beam away from the hard target to obtain atmospheric backscattering, **Figure 7.**

### 4.6 Real-time wind measurement

In this section, real-time wind measurements are reported. Real-time measurements are recoded continuously, thanks to FPGA pre-processing techniques that allow for streaming of either power spectrum or autocorrelation of received signals instead of raw data. The instrument was installed in our research vehicle that is located at the City College of New York at upper Manhattan, New York (latitude: 40.49°N, longitude: 73.56°W). Laser pulses are transmitted into the atmosphere through an opening in the vehicle's roof, **Figure 8**. The following subsections introduce vertical wind velocity measurements using FFT and autocorrelation pre-processing techniques. Horizontal wind speed measurement is also introduced.

### 4.7 Vertical wind velocity measurement using FFT pre-processing algorithm

Received signals are pre-processed on the FPGA by dividing backscattered signals into 128 samples per a range gate, corresponding to a 48 m range resolution,

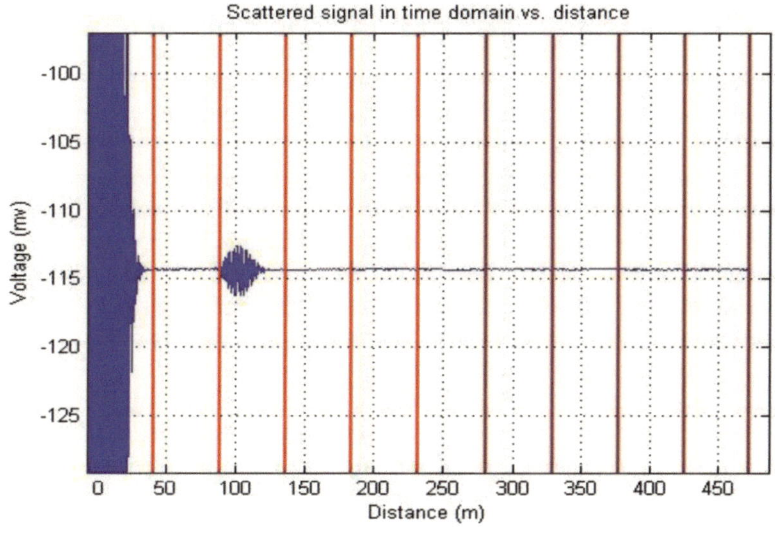


**Figure 6.**
*Time domain scattered signal off a hard target.*

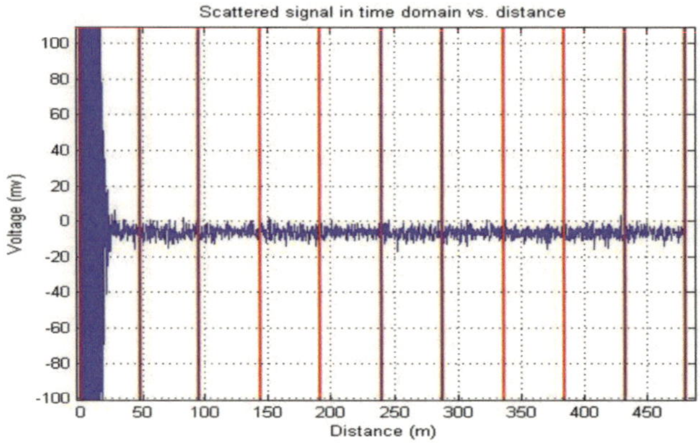


**Figure 7.**
*Time domain scattered signal off the atmosphere.*

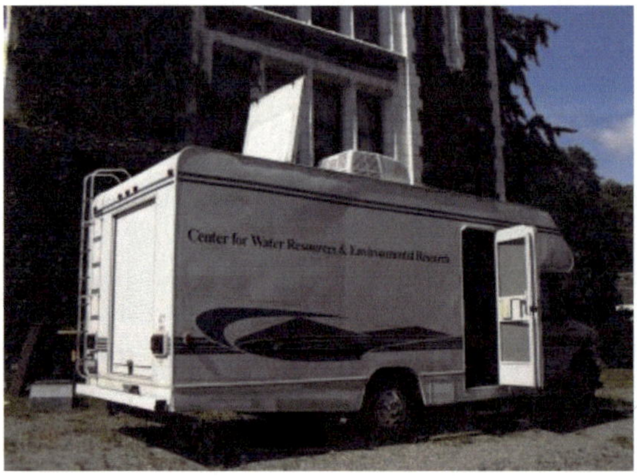

**Figure 8.**
*The research van where the Doppler lidar instrument is installed for field measurements.*

ranging from 96 m to approximately 3 km. Power spectra of the gated signals are estimated by calculating the FFT and then accumulating the resulted power spectrum for 10,000 shots. Accumulated power spectra are streamed to the host PC for further processing and archiving. Wind velocities are calculated by estimating the Doppler frequency shift of the received signals' power spectrum. Three different techniques are used for velocity estimate; spectral peak search, power spectrum peak curve fitting, and MLE algorithm.

In the following section, vertical measurement results of wind velocity measured during the month of August of 2011 are introduced and discussed. The vertical wind velocity profile was measured on August 17th, 2011 from 14:35-to-16:35 EDT. Scattered signals' power spectrum is flattened by dividing it by the power spectrum of input signals to the photo-detector when no scattered signals are present. A signal intensity threshold is chosen below which any returned signals power spectrum is ignored. In **Figure 9**, vertical wind velocity is estimated by fitting the power spectrum profile around the peak value to a Gaussian curve shape and finding the centroid of the fitted shape. It is also shown that updrafts and downdrafts are very visible, which is typical weather condition during a summer afternoon.

Backscattered signals' power was estimated by integrating the area under the power spectrum curve, **Figure 10**. The signal power was then range corrected by dividing estimated power by the overlap integral function given that the laser beam was focused at a known distance (approximately 2 km), which was determined by scattering off a hard target (clouds).

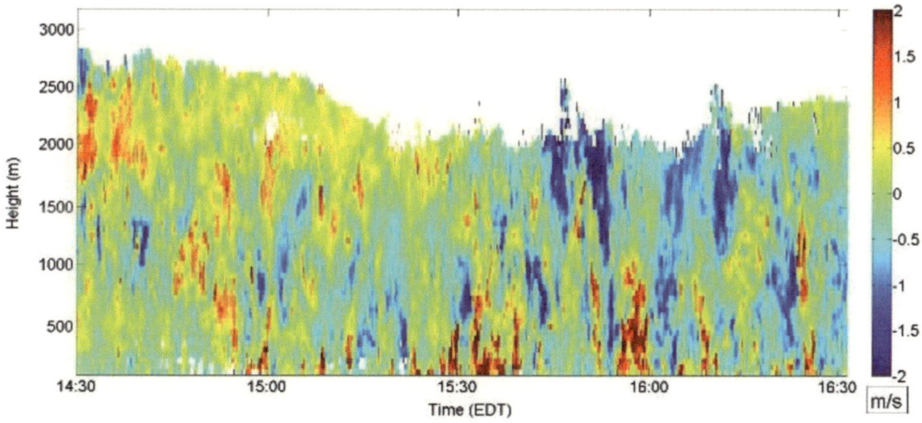


**Figure 9.**
*Vertical wind velocity profile (m/s) measured at CCNY on August, 17th, 2011 from 14:35–16:35 PM EDT estimated by Gaussian curve fitting.*

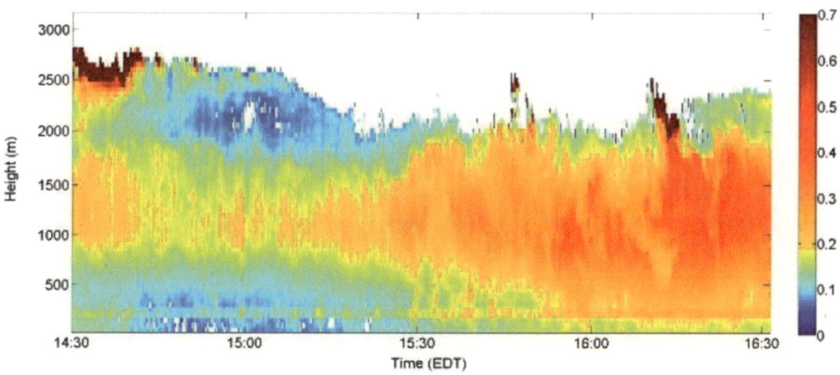


**Figure 10.**
*Ratio of received signals power to shot noise power V.s. time and height measured at CCNY on August, 17th, 2011 from 14:35–16:35 PM EDT.*

Range correcting received signals' power allows for comparison our results with the results of a 1 μm direct detection lidar, which was operated at the remote sensing laboratory of the City College of New York during the same time of operation. **Figure 11A** and **B** show a good agreement of signal intensity profile's and cloud patterns with that of the CDL at 14:30, 15:50 and 16:15. It is also noticed that signal intensity increased significantly at a height of approximately 2700 m at 14:30 and 16:20 due to clouds at that height. Both lidars also show a gradual increase of aerosol concentration as a function of time between 16:00 and 16:30. The Doppler lidar's signals power spectrum comparison with the direct detection measurements proves that the obtained power spectrum is valid, and as a result, the frequency shift has to be due to: (a) RF signals going into the AOMs, and (b) Doppler shift caused by wind. Since the LO is stable (laser's line width is approximately 3 kHz), and the jitter between the RF signal and the ADC's clock is less than 100 kHz, then measured frequency shifts are solely due to Doppler shifts caused by wind velocity.

As mentioned in Section 2, the power spectrum of received signals needs to be corrected for the detector's non-flat gain shape. This is done by dividing the power spectrum of received signals by a power spectrum of reference signals. The reference signals are obtained using the detector while LO is turned on and no received signals are present (final amplifier is turned off), i.e. this power spectrum only represents shot noise. The SNR can be represented as follows:

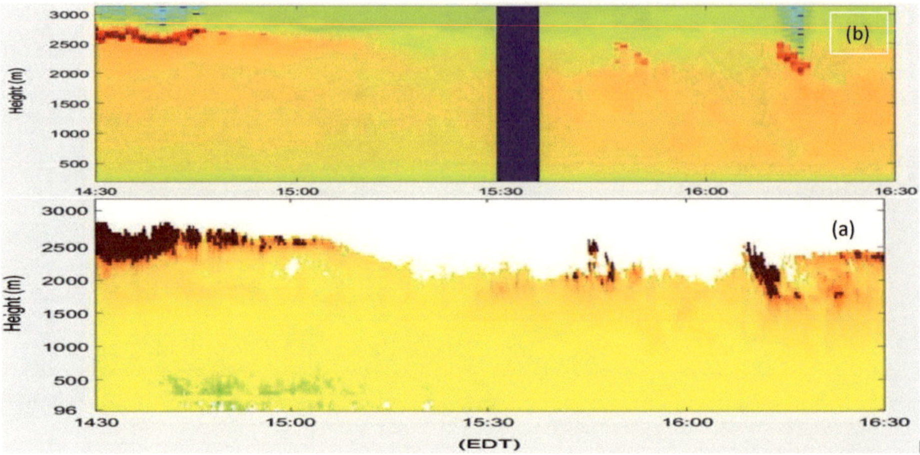


**Figure 11.**
*Range corrected backscattered signal power V.s. time and height (a) and the 1 μm direct detection lidar signal power vs. height and time (b). Both signals power profiles show a good agreement around 14:35, 15:60, and 16:15, where clouds' patterns are observed at the same heights. Aerosols concentration profiles also show a good agreement in the two measurements.*

$$SNR = \frac{P_{sig}^{(N)}}{P_{nois}} \tag{32}$$

where, $P_{nois}$, and $P_{sig}^{(N)}$ are the noise and signal power accumulated over N pulses, respectively. For a shot noise limited receiver system, the noise power is due to shot noise: $P_{shot}^{(N)}$, which can be given by:

$$P_{shot}^{(N)} = <P_{shot}> \pm \sigma_{shot}^{(N)} \tag{33}$$

where; $<P_{shot}>$ is the average shot noise power, which is a fixed level characterized by the laser source, and $\sigma_{shot}^{(N)}$ is the shot noise variation around its fixed level when averaged N times (standard deviation of the shot noise for N accumulation). The average shot noise fixed power level is given by:

$$P_{shot} = 2eiB \tag{34}$$

where; $e$ is the electronic charge, $i$ is the photo-current, and $B$ is the detector's bandwidth.

In our analysis, we calculate the following parameter: $\frac{P_{measured}}{P_{ref}} - 1$ by dividing signals power spectrum by a reference signals power spectrum and then subtracting one, which results to:

$$\frac{P_{measured}}{P_{ref}} - 1 = \frac{(SNR)}{\sqrt{N}} + \sqrt{\frac{2}{N}} \tag{35}$$

Therefore, when accumulating 10,000 pulses and in order to extract signal out of noise, the signal power should be at least equal to noise power, i.e. SNR =1. As a result, the value we calculate is equal to: $\frac{1}{\sqrt{10,000}} + \frac{\sqrt{2}}{\sqrt{10,000}}$ = 0.024, which is the value we set as a threshold below which received signals are ignored. It is also worth noting that the signal power we report is not really the signal power, but it's the signal power normalized to the shot noise, in other words, it is the SNR for a single shot.

## 5. FPGA programming and wind measurements analyzed using autocorrelation

The objective of processing received signals using this technique is to have the ability to change the spatial resolution. This is achieved by calculating the autocorrelation of received signals and using it to calculate the power spectrum of any desired range gate [24]. The power spectrum of received signals is found by calculating the FFT of the autocorrelation as shown in Eqs. (36) and (37).

$$R(\tau) = \int_{-\infty}^{\infty} f(t)f(t+\tau)dt \tag{36}$$

$$G(f) = \int_{-\infty}^{\infty} R(\tau)e^{-j2\pi f\tau}\,dt \tag{37}$$

where; $f(t)$ is a time domain signal, $R(\tau)$ is the signal's autocorrelation, and $G(f)$ is the Fourier transform (power spectrum).

Changing range gates (varying spatial resolution) is an advantage that previous FFT pre-processing algorithm does not have. In this technique (autocorrelation), digitized received signals are split into two paths. The first path is mixed with a cosine signal oscillating at 84 MHz to produce an in-phase (I) component; the other path is mixed with a sine signal oscillating at 84 MHz to produce a quadrature (Q) component, **Figure 12**.

### 5.1 Autocorrelation (analog complex demodulator) pre-processing algorithm

Mixing the received signals (oscillating around 84 MHz +/− Doppler shift) with an 84 MHz cosine and sine waves produces two output signals; a high frequency (sum of the two frequencies) component and low frequency (difference of the two frequencies) component (the Doppler shift). A low-pass, finite impulse response (FIR), filter is used on each path to get rid of the unwanted high frequency. Filtered signals are then down-sampled (decimated) by a factor of 4, which will reduce our original sampling period from 2.5 n.s (400 MHz) to 10 n.s (100 MHz). This down conversion reduces the maximum detectable frequency (according to Nyquest theorem) to 50 MHz, which corresponds to a radial velocity of approximately 38 m/s. The resulting complex time sequence $d(n) = d_i(n) + j\,d_q(n)$ is input to the (M)-lag autocorrelator circuit,

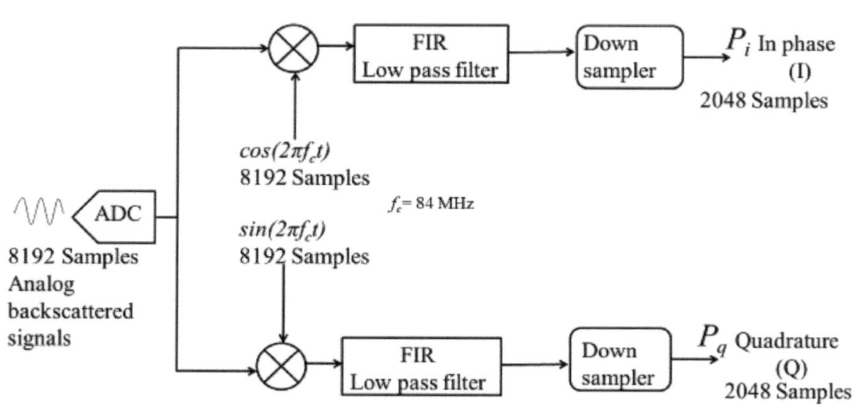


**Figure 12.**
*Autocorrelation algorithm block diagram as implemented on the FPGA to produce an in-phase (I) and a quadrature (Q) signals.*

which computes an autocorrelation matrix $D(m,n) = d^*(n).d(n+m)$ for m = 0 to M-1 (lags) and n = 0 to N-1 (number of time domain samples, which is 8k samples/4 = 2 k), where $d^* = d_i(n) - j\, d_q(n)$ is the complex conjugate of $d(n)$. The processing repeats for 10 k laser shots and the elements of D matrix are accumulated and then streamed to an output buffer before it is being streamed to the host PC.

Once the accumulated lags' matrix [Eq. (5)–(3)] is streamed to the host PC, further processing is conducted to calculate the power spectrum of received signals as follows:

$$D = \begin{bmatrix} S_0S_0^* & S_0S_1^* & S_0S_2^* & \cdots & \cdots & S_0S_{M-1}^* \\ S_1S_1^* & S_1S_2^* & S_1S_3^* & \cdots & \cdots & \vdots \\ S_2S_2^* & S_2S_3^* & S_2S_4^* & \cdots & \cdots & \vdots \\ \vdots & \vdots & \vdots & \cdots & \cdots & \vdots \\ S_{n-1}S_{n-1}^* & S_{n-1}S_{n-2}^* & 0 & \cdots & \cdots & \vdots \\ S_nS_n^* & 0 & 0 & \cdots & \cdots & \vdots \end{bmatrix} \quad (38)$$

where; $M$ is the number of lags, $n$ is the number of acquired samples, $S$ denotes to a sample, and $S^*$ denotes to the complex conjugate of sample $S$.

To calculate the power spectrum of a certain range gate, the columns of the D matrix are accumulated from the $i^{th}$ row to the $j^{th}$ row, where $i$ and $j$ are the first and last corresponding samples of that range gate, respectively. This accumulation process produces an M size autocorrelation vector, which is complex (in-phase and quadrature components). Since the autocorrelation is symmetric, we construct the second half of the autocorrelation vector by making its real part even and imaginary part odd. Finally, we find the power spectrum of that range gate's signals by calculating the FFT of the constructed complex autocorrelation vector.

## 5.2 Vertical wind velocity measurements using autocorrelation pre-processing algorithm

In this section, wind velocity was measured in a vertical mode while pre-processing received signals using an autocorrelation algorithm. The autocorrelation algorithm calculates the autocorrelation of the received signals and streams out the lags matrix that can be gated according to user's range resolution's preference. That feature makes autocorrelation technique advantageous over the FFT technique, where range gates are fixed. Wind velocity was measured under this mode of operation on July 12th, 2012, **Figure 13**.

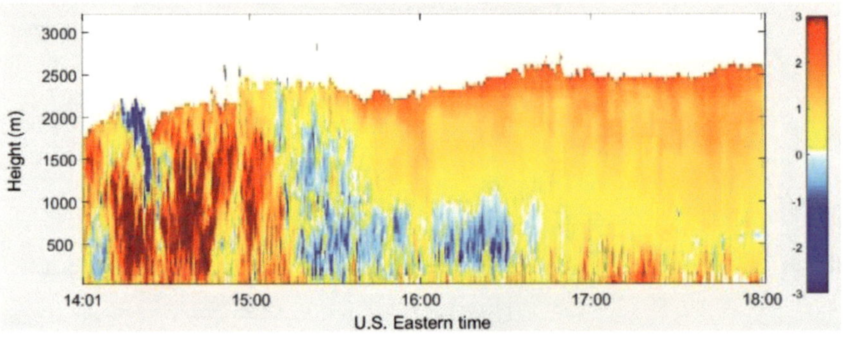


**Figure 13.**
*Vertical wind velocity (m/s) vs. time and height measured at CCNY remote sensing laboratory between 14:01–18:00 PM EDT on July 12th, 2012.*

## 6. Conclusion

In conclusion, an eye-safe all-fiber CDL system for wind sensing in urban areas was designed, developed, tested, and operated at the remote sensing Laboratory of the City College of New York.

The system utilizes a 1.5 μm fiber optics laser, which benefits from the availability and affordability of telecommunication optical components. Two AOMs are connected in series to achieve a high extinction ratio and to shift the laser frequency by 42 MHz each, which produces a total shift of 84 MHz. An optical amplifier amplifies the laser pulse to produce approximately 12 μJ/pulse (200 ns FWHM at 20 kHz PFR). An optical circulator directs amplified laser pulses to its output port that is connected to the optical antenna, and directs received signals to an optical coupler to be mixed with a LO. Circulator's fiber tip was polished and angled to reduce internal reflection that can damage the detector. Optical mixed signals are detected by a heterodyne balanced detector.

Received signals are sampled at 400 MHz through a 14-bit ADC equipped with an FPGA. Due to the very low energy per pulse (12 μJ/pulse), a high PFR (20 kHz) is used to allow for digging the very low signal out of noise. This high pulse rate makes it almost impossible to process the data in real time, therefore, the FPGA was programmed to pre-process received signals at the hardware level as the received signals are being acquired and before streaming to the host PC.

Two different pre-processing algorithms have been simulated and programmed into the FPGA; one algorithm calculates FFT of time gated received signals and accumulates the resulted power spectrum; the other algorithm calculates autocorrelation of the received signals and accumulates the result. The later algorithm allows for changing range gate (spatial resolution), which can be applied to signals scattered from very high altitudes (where signals are very weak) to improve the SNR.

The system was installed in a research vehicle and wind velocity was measured at the City College of New York. Wind velocity was measured in two different modes; vertical mode, and scan mode. Wind velocity was measured up to 3 km in a vertical mode during a very clear day. The system can be operated to measure wind velocity, processes received signals in real time, and display results while acquiring data. Improving the system can be achieved by increasing the measured range to 7 km instead of 3 km.

## Author details

Sameh Abdelazim[1]*, David Santoro[2], Mark F. Arend[2], Fred Moshary[2] and Sam Ahmed[2]

1 Fairleigh Dickinson University, New Jersey, USA

2 The City College of New York, New York, USA

*Address all correspondence to: azim@fdu.edu

## References

[1] Huffaker RM, Jelalian AV, Thompson JAL. Laser-Doppler system for detection of aircraft trailing vortices. IEEE. 1970;**58**:322-326

[2] Kameyama S, Ando T, Asaka K, Hirano Y, Wadaka S. Compact all-fiber pulsed coherent Doppler lidar system for wind sensing. Applied Optics. 2007; **46**:1953-1962

[3] Karlsson CJ, Olsson FÅA, Letalick D, Harris M. Ll-fiber multifunction continuous-wave coherent laser radar at 1.55 νm for range, speed, vibration, and wind measurements. Applied Optics. 2000;**39**:3716-3726

[4] Henderson SW, Suni PJ, Hale CP, Hannon SM, Magee JR, Bruns DL, et al. Coherent laser radar at 2 mm using solid state lasers. IEEE Transactions on Geoscience and Remote Sensing. 1993; **31**:4-15

[5] Kane TJ, Kmetec JD, Wagener TJ. Flight test of 2-mm diode pumped laser radar system. In: SPIE 2464, Air Traffic Control Technologies. Orlando, FL; 1995

[6] Kavaya MJ, Henderson SW, Magee JR, Hale CP, Huffaker RM. Remote wind profiling with a solid-state Nd:YAG coherent lidar system. Optics Letters. 1989;**14**:776-778

[7] Abdelazim S, Santoro D, Arend M, Moshary F, Ahmed S. Development and operational analysis of an all-fiber coherent Doppler lidar system for wind sensing and aerosol profiling. IEEE Transactions on Geoscience and Remote Sensing. 2015;**53**(12):6495-6506

[8] Arend M, Santoro D, Abdelazim S, Gross B, Moshary F, Ahmed S. Development of a NYC Meteorological Network with Emphasis on Vertical Wind Profiles in Support of Meteorological and Dispersion Models. Phoenix, AZ: American Meteorological Society; 2009

[9] Cariou J-P, Augere B, Valla M. Laser source requirements for coherent lidars based on fiber technology. Comptes Rendus Physique. 2006;**7**:213-223

[10] Holmes FJ, Rask BJ. Optimum optical local-oscillator power levels for coherent detection with photodiodes. Applied Optics. 1995;**34**:927-933

[11] Amzajerdian F, Pierrottet D, Singh U, Kavaya M. Optimum integrated heterodyne photoreceiver for coherent lidar applications. MRS. 2005;**883**: PROC-883-FF6.3

[12] Darcie T, Moye A, Driessen P, Bull J, Kato H, Jaeger N. Noise reduction in class-AB microwave-photonic links. Microwave Photonics. 2005;**12**(14): 329-332

[13] Frehlich RG, Kavaya MJ. Coherent laser radar performance for general atmospheric refractive turbulence. Applied Optics. 1991;**30**:5325-5352

[14] Clifford SF, Wandzura S. Monostatic heterodyne lidar performance: The effect of the turbulent atmosphere. Applied Optics. 1981;**20**: 514-516

[15] Wandzura SM. Meaning of quadratic structure functions. Journal of the Optical Society of America. 1980;**70**: 745-747

[16] Kavaya MJ, Menzies RT, Haner DA, Oppenheim UP, Flamant PH. Target reflectance measurements for calibration of lidar atmospheric backscatter. Applied Optics. 1983;**22**: 2619-2628

[17] Yanzeng Z, Post MJ, Hardesty RM. Receiving efficiency of monostatic

pulsed coherent lidars. 2: Applications. Applied Optics. 1990;**29**:4120-4132

[18] Fried DL. Optical heterodyne detection of an atmospherically distorted signal wave front. IEEE. 1967; **55**:57-66

[19] Kameyama S, Ando T, Asaka K, Hirano Y, Wadaka S. Performance of discrete-Fourier-transform-based velocity estimators for a wind-sensing coherent Doppler lidar system in the Kolmogorov turbulence regime. IEEE Transactions on Geoscience and Remote Sensing. 2009;**47**(10):3560-3569

[20] Hardesty RM. Performance of a spectral peak frequency estimator for Doppler wind velocity measurement. IEEE Transactions on Geoscience and Remote Sensing. 1986;**GRS-24**(5): 777-783

[21] Doviak RJ, Zrnic DS. Doppler Radar and Weather Observations. Orlando, FL: Academic; 1984

[22] Hardesty RM. Performance of a discrete spectral peak frequency estimator for Doppler wind velocity measurements. IEEE Transactions on Geoscience and Remote Sensing. 1986; **GE-24**(5):777-783

[23] Sirmans D, Bumgarner B. Numerical comparison of five mean frequency estimators. Meteorological Applications. 1975;**14**:991-1003

[24] Abdelazim S, Santoro D, Arend M, Moshary F, Ahmed S. A hardware implemented autocorrelation technique for estimating power SPectral density for processing signals from a Doppler wind Lidar system. Sensors. 2018;**18** (12):4170. DOI: 10.3390/s18124170

Chapter 2

# Low-Key Stationary and Mobile Tools for Probing the Atmospheric UHI Effect

*Kristen Koch, Gunnar W. Schade, Anthony M. Filippi, Garrison Goessler and Burak Güneralp*


## Abstract

The urban heat island (UHI) effect is created by a series of man-made surface modifications in urban areas that cause changes to the surface energy balance, resulting in higher urban surface air temperatures as compared with surrounding rural areas. Studying the UHI effect is highly amenable to hands-on undergraduate student research projects, because, among other reasons, there are low key measurement tools that allow accurate and regular stationary and mobile probing of air temperature. Here, we summarize the results of a student project at Texas A&M University that analyzed the atmospheric UHI of Bryan/College Station, a mid-size metro area in east Texas. Sling psychrometers were used for semi-regular twice daily stationary air temperature monitoring, and a low-cost electronic sensor and miniature data logger were used for mobile measurements. Stationary data from two similar, open mid-rise building locations showed typical UHI intensities of 0–2°C, while the mobile measurements identified situations with UHI intensities exceeding 6°C when traversing areas with high impervious surface fractions. Nighttime measurements showed the expected UHI intensity relations to wind speed and atmospheric pressure, while daytime data were more strongly related to urban morphology. The success of this research may encourage similar student projects that deliver baseline data to urban communities seeking to mitigate the UHI.


**Keywords:** sling psychrometer, student project, UHI intensity, mobile measurements, impervious area fraction

## 1. Introduction

The urban heat island (UHI) effect is a relatively well-researched surface meteorological phenomenon. It describes the difference in surface air temperature between a built-up urban area and its surrounding countryside. Said difference is usually displaying a warmer surface layer air temperature in urban areas, especially during nighttime. Several reviews of the UHI effect magnitudes, characteristics, causes and mitigation strategies can be found in the literature [1–7]. Arnfield [1] summarized the major aspects, stating that UHI intensity is highest at night, typically increases with lower wind speeds under clear sky, high pressure conditions, and is usually more pronounced during summer time and in larger, more populous cities. Recent critique and recommendations for ongoing research [7, 8] have led to a more streamlined approach of interpreting UHI intensity on the basis of measurement techniques, locations, and urban morphological characterizations such as impervious area fractions, building heights, and canyon aspect ratios.

The causes of UHI effects are related to fundamental differences in the surface energy balance between urban and rural areas. The 3D structure of, and man-made materials in, urban areas cause albedo changes during daytime and "radiation trapping" at night [6, 9–13], causing stronger heat admission during daytime, and slower radiative heat losses at night. In addition, anthropogenic heat from the human population and its energy use in urban areas significantly enhances the UHI effect [14–22]. While rural areas convert a substantial fraction of daytime incoming net radiation into latent heat fluxes, the dominance of impervious areas and an associated lower vegetation density in urban areas compared with their rural surroundings causes a redistribution of incoming net radiation into urban heat storage and sensible heat fluxes. Increased sensible heat fluxes increase the daytime UHI intensity, while high heat storage fluxes exacerbate nighttime UHI intensities when stored heat is returned into the atmosphere [23–28]. Detailed numerical studies such as by Ryu and Baik [29] have shown that impervious surface area, a proxy for energy balance flux changes, is likely the dominant factor determining daytime UHI intensity, while anthropogenic heat releases may dominate nighttime UHI intensity. Both these factors interact with the 3D structure of the urban fabric and the prevailing meteorological conditions. This can cause daytime cool islands as man-made (impervious) surfaces store heat and can shade road "canyons"; and maximum nighttime heat islands as stored heat together with anthropogenic heat are released back into shallower nighttime surface air layers. The results also concur with higher net radiation levels under high pressure conditions in summer, and the associated lack of turbulent heat transport under low wind speeds in urban areas as summarized by Arnfield [1].

To investigate these phenomena, researchers have used both stationary and mobile air temperature measurements extensively. While early studies often used only a few weather station locations [30, 31], or limited mobile traverses [32–34], newer studies have profited from now widely available, small form factor, accurate, and cost-effective electronic temperature sensors deployed in either stationary or mobile fashion. However, the correct deployment and interpretation of such sensors and their data still requires careful consideration, such as of radiation shielding and sensor response time aspects. In comparison, a hand-operated sling psychrometer provides a highly accurate, battery-independent low-key tool that can be operated by any lay person and can be immediately ready at the required time. Sling psychrometers provide dry-bulb and wet-bulb temperatures, and thus serve to provide both air temperature and humidity. They have been used in the past for UHI "spot" measurements [35–37], supplementing weather station and mobile data, and are ideally suited as "hands-on" data collection tools in undergraduate student research projects [38].

This chapter describes a semester-long student project to determine the UHI intensity of a mid-size metropolitan area in east Texas, the Bryan/College Station (BCS) metro area, home of Texas A&M University. As part of a spring semester course on environmental atmospheric science, students were tasked to maintain regular air temperature measurements near the places they lived in town, then turn in a writing assignment at the end of the semester. During the following summer and fall semesters, the first author maintained two of the measurement sites and also carried out a mobile measurement study using her private automobile. Here, we discuss selected results from the measurements in context of past UHI studies. We also introduce an ongoing project of integrating these measurement results with remotely sensed land cover data.

## 2. Methods

### 2.1 Sling psychrometer measurements

Sling psychrometers are traditional meteorological measurement tools to determine air temperature (dry-bulb temperature) and relative humidity (wet-bulb temperature). Sling psychrometers have been used in UHI studies going back several decades [32, 35, 36, 39]. They have an educational advantage over automated measurements as they require direct student involvement in the data gathering and documentation process. The instrument used in this study was a Bacharach model 0012-7012 using two red spirit filled glass thermometers with Fahrenheit scales. The instrument is made of a robust hard plastic shell, with its outer part acting as handle when extended, while the inner part bears the two identical glass thermometers. The thermometer scales allow readings as precise as 0.5°F, and are accurate to at least 1.0°F based on intercomparisons with other sling psychrometers and a research grade meteorological sensor, comparisons that were made part of the student project in this UHI study.

Here, we discuss only the dry bulb, aka air temperature data. The atmospheric UHI effect was calculated as the difference between the measured air temperature and the corresponding temperature at 2 m above ground level (agl) at the Texas A&M weather station (**Figure 1**). The weather station's combined T/RH sensor, a

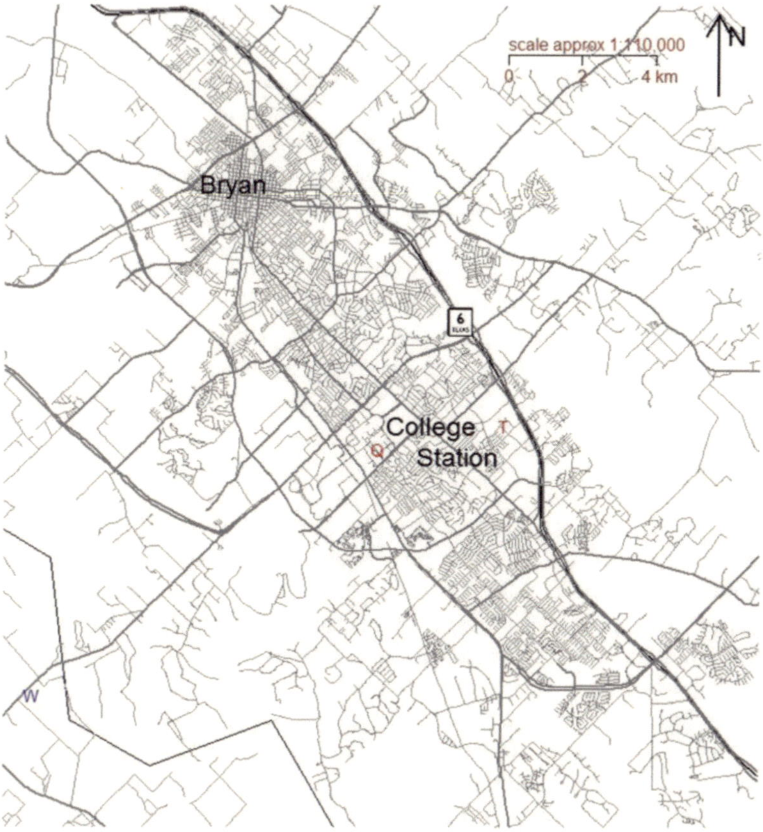


**Figure 1.**
*Roadmap based view of the Bryan/College Station metro area in East Texas (30.6°N, 96.32 W). The major highway traversing the area, Texas-6, is labeled alongside the two stationary measurement locations in red (T = 'trails', Q = 'quad') and the weather station location (W) in blue.*

model 085 from MetOne Instruments Inc., has a high precision and accuracy multi-element thermistor (±0.15°C). Corrections for elevation differences between the measurement sites and the weather station reference location were made using the dry adiabatic lapse rate.

Multiple locations were monitored within the BCS metro area during the spring semester of 2015, but only two urban locations, which were monitored throughout 2015, are discussed here. The first location, subsequently identified as the "Trails" apartment complex, was a parking lot area in front of apartment buildings (**Figure 1**) in College Station. This part of town is slightly elevated from the surrounding urban area, and located between a narrow, wooded green corridor, and a major freeway (State Highway 6). While the location was next to pervious lawns, the area was almost tree-less and dominated by buildings, roads, and parking lots. The tallest buildings are the three story apartment buildings; the larger area within half a kilometer, however, includes a midsize mall and associated parking lots to the north, commercial buildings, including hotels, along the freeway to the east, and both taller and less tall apartment buildings toward the south and west.

The second location, subsequently referred to as "Quad", was a small parking lot on the Texas A&M campus in College Station (**Figure 1**). This location is also dominated by impervious surface areas, such as more extensive parking lots to the southeast, large parking garages to the east, and the onsite multi-story dormitories and Dining Hall. It has, however, numerous trees lining the nearby streets. The surroundings within half a kilometer consist of numerous, multi-story university buildings to the north and west, an open park, a field, and a wooded, one-story residential neighborhood to the southwest and south, and a golf course further to the northeast.

Both locations can be characterized as local climate zones (LCZ) $5_6$, open mid-rise to low-rise urban areas with 30–50% impervious area [8]. However, Trails is located closer to the east end of town, with rural areas (LCZ B/C) as close as 1.5 km to the east, while Quad is located more central to the metro area, with open, sparsely build-up LCZ 9 areas more than 2 km distant, and rural areas more than 4 km distant (**Figure 1**).

UHI intensity was determined for each location by comparing observed local temperatures to the rural temperature measurement (10-min average) at Texas A&M's weather station 10 km outside the urban area toward the southwest. In addition, we assembled weather station data (pressure, winds, solar radiation, and precipitation) for the days measurements were taken for data analysis.

### 2.2 Mobile measurements

Air temperature sensors and associated logging equipment have become both miniaturized and highly affordable, thus allowing both high spatial density distributed and low-key mobile data collection of the urban heat island [40]. Mobile measurements (traverses) using automobiles, but also bicycle-mounted or personally carried sensors, have been used numerous times in the past to study the UHI effect, both in long- and in short-term campaigns [32, 33, 37, 41–65].

In summer and fall 2015, we carried out an undergraduate student research project [38] using a standard HOBO U12 data logger with 12-bit temperature sensor (model TMC6-HD) from Onset Computer Inc. to evaluate the atmospheric urban heat island of the BCS metro area. The sensor has a typical accuracy of ±0.25°C and is recorded at a better than 0.2°C resolution. It was placed into a passive radiation shield (model RS3, Onset Comp. Inc.), which was mounted to an angular steel bracket attached to a metal sleeve with a flat rubber sheet that slides over a passenger car side window (**Figure 2**). With the window moved nearly all the way up, the

**Figure 2.**
*Mobile measurement setup.*

sensor was measuring air temperature at approximately 2 m agl off the passenger side of the car, with its logger safely placed inside the air-conditioned vehicle. This avoided most temperature bias effects from the vehicle itself, or from other vehicles on the road, unless measurements occurred either at very low speeds with air moving across the front hood of the car toward the sensor, such as may have occurred when stopping at traffic lights; or when hot exhaust plumes from vehicles on the road were encountered. Such possible temperature measurement biases, though occasionally encountered, were not removed from the data set.

Typical car moving speeds were 20–30 mph in town and up 50 mph outside the urban areas (approximately 10–25 m $s^{-1}$) during all mobile measurements. A smart phone app called RAAH (https://www.raah.co) was used to record the vehicle's location. Data logger and smart-phone times were aligned before each drive. During each traverse data were recorded every 10 s, providing for a typical horizontal resolution of 100+ m. However, considering the sensor's response time, and the accuracy and timing of the location determination, both the absolute bias and uncertainty of a recorded temperature's location were likely larger than 100 m. Such aspects should be considered in all mobile studies when moving speeds are comparable or faster than sensor response times.

Raw temperature data were normalized first for the presumed linear temperature change, if present, during the typically one-hour drive for data collection, then adjusted for elevation differences along the route using the dry adiabatic lapse rate. Both corrections were always significantly smaller than the encountered temperature differences between the rural and (sub-)urban areas traversed.

The routes driven were selected to avoid selective coverage of the metro area, and such that measurements along a route could be completed within approximately 1 h, including nearby rural areas. All routes were driven once in the morning, and then at least two times during the afternoon and around midnight. Based on student availability during the project, completing the set of selected routes lasted from late August to early November 2015.

One route, which was driven twice during the project, brought the vehicle close to the Texas A&M weather station in the rural area approximately 10 km outside the urban area to the southwest (**Figure 1**). Good agreement between the vehicle based sensor and the weather station sensor was observed (<0.5°C differences), and minimum temperatures during each drive were strongly correlated with weather station temperature during the drive hour (slope = 0.98, $r^2 > 0.99$). Consequently, we used calculated temperature anomalies during each traverse and determined UHI intensity from each drive's observed maximum minus minimum temperature. The remaining meteorological parameters were assembled in similar fashion as described above for the stationary measurements.

## 2.3 Urban area characterization using remote sensing

To aid in the interpretation of the observed atmospheric UHI effect, we estimated the spatial extent and configuration of various land-cover/land-use (LCLU) areas in Bryan/College Station (BCS), Texas. To this end, we jointly analyzed airborne digital orthophotographs and a LiDAR-derived digital surface model (DSM) of the study area. We thus characterized urban LCLU based on a remote-sensing approach. In particular, we employed a geospatial object-based image analysis (GEOBIA) method [66, 67], which effectively exploits contextual/spatial information. GEOBIA-based image-processing algorithms and data fusion are needed to more fully exploit image information, particularly in the case of high-spatial-resolution data. For our GEOBIA analysis, we employed eCognition® Developer software, which allows objects/segments to be delineated and used to classify geospatial datasets. We divided the processing steps into two phases: an initial remote-sensing data segmentation and classification phase, and a subsequent post-segmentation/classification editing phase, which we conducted to improve classification accuracy. We performed quantitative classification accuracy assessment on the revised, post-classification edited result.

### *2.3.1 Segmentation and classification phase*

*Data sets and data pre-processing.* Data processing for this remote-sensing segmentation/classification analysis began with the acquisition of multiple 50-cm natural-color (NC)/color-infrared (CIR) Digital Orthophoto Quarter Quad (DOQQ) images, acquired from the Texas Natural Resource Information System (TNRIS). These images were collected between October 2014 and August 2015 as part of the 2015 Texas Orthoimagery Program. We mosaicked the DOQQs for the BCS, Texas study area, and then spatially resampled the mosaic to a 5-m grid cell size in order to facilitate subsequent completion of GEOBIA computational processing. Additionally, we utilized a high-point-density LiDAR point cloud, acquired from Texas A&M University, which was collected in conjunction with the State of Texas over the February 9–10, 2015 period. We filtered and processed the LiDAR point cloud using Esri ArcGIS LAS tools to produce a digital surface model (DSM) of the BCS area. After initial processing of the image and LiDAR data, we spatially subset the respective data sets using a polygon to fit the same areal extents.

*Information classes and training set delineation.* The LCLU information classes employed in this analysis are: roads, roofs, other impervious, trees, lawns, water, bare soil, and pasture/cropland. The pasture/cropland class is actually a combined pasture/grassland/cropland class, which contains grasslands that are not (residential)

| Class ID number | Class name | Number of training polygons |
|---|---|---|
| 1 | Roads | 62 |
| 2 | Roofs | 281 |
| 3 | Other Impervious | 43 |
| 4 | Trees | 43 |
| 5 | Lawns | 153 |
| 6 | Water | 219 |
| 7 | Bare Soil | 39 |
| 8 | Pasture/Cropland | 46 |

**Table 1.**
*Number of training polygons per class, used for GEOBIA-based LCLU classification of Bryan/College Station (BCS), Texas study area.*

lawns. We manually delineated the training areas based on manual/visual interpretation of the aerial photography and DSM. We initially generated training areas for a set of 15 classes, then later merged some of these classes, yielding the aforementioned final set of eight information classes. The distribution of training-area polygons across these classes is given in **Table 1**.

*GEOBIA parameter values, input variables, and ruleset development*. We created a ruleset within eCognition® Developer using the imagery, DSM, and training areas. The process involved multiple steps, including: segmentation, class assignment using the training areas, assigning the classified segments as samples, configuring the nearest-neighbor classifier, and applying/conducting the classification. We determined GEOBIA parameter values via iterative, trial-and-error experimentation, with the objective of maximizing LCLU map classification accuracy, while also accounting for computational constraints of the GEOBIA system/computing environment. For the classification step, we used a nearest-neighbor classification algorithm. The classification input features included: mean blue band, mean green band, mean red band, mean near-infrared (NIR) band, mean DSM, standard deviation blue band, standard deviation green band, standard deviation red band, standard deviation NIR band, standard deviation DSM, normalized difference vegetation index (NDVI) [68], and normalized difference water index (NDWI) [69]. We applied the classifier at the image-object level [66], and exported classification results in both vector and raster formats, where we subjected the latter product (at 5-m cell size) to subsequent processing and analysis.

### 2.3.2 Post-segmentation/classification editing phase

The resultant LCLU classification entailed various areas of clear misclassifications, including some roads, parking lots, and roofs/buildings. Therefore, we performed some post-/segmentation/classification editing to increase classification accuracy of the class information. In particular, publicly available geographic information system (GIS) data from the City of College Station and the Brazos County Appraisal District aided editing of the LCLU classification map. For roads, we used a vector GIS layer containing road center lines, where we buffered the center lines, 5 m on each side, followed by a dissolve operation. We assigned the road classification value to the buffered area, converted to the data to the raster data model, and then overwrote the pixels in the original LCLU map in these areas with these road classification values. For the roofs class, we followed a similar procedure, except that no buffers were used. Since we only had access to building footprint data from the City of College Station, we only applied this operation to areas within College Station, and not the City of Bryan; however, a large portion of the study area lies within the City of College Station boundary. For the other impervious class, based on visual interpretation of the aerial photography, we digitized impervious areas—especially parking lots—where misclassifications were evident. After rasterization of the digitized polygons, we then overwrote misclassified pixels in the original LCLU raster using these newly digitized data. Based on visual inspection, these post-classification GIS operations generally markedly improved classification accuracy. One caveat though is that although the DOQQs and LiDAR point clouds were collected close in time, the data in the vector GIS layers used in this post-segmentation/classification editing phase were acquired some time prior to the DOQQ and LiDAR acquisitions, and this temporal disjunction may have resulted in some errors in the edited LCLU map, particularly given the relatively high rate of urbanization in the BCS area in recent years. Furthermore, future analysis may entail using variable buffer distances for the road centerlines, depending upon road type.

*2.3.3 Quantitative land-cover/land-use classification accuracy assessment*

Using standard methods [70, 71], we performed a quantitative thematic/classification accuracy assessment on the finalized land-cover/land-use (LCLU) classified map, edited post-classification. We employed a stratified random sampling approach for generating the classification accuracy-assessment points/sample locations, and we used 100 such points per class [70, 71]. We used the DOQQ aerial photograph mosaic as the reference data; we evaluated the accuracy of the post-classification-edited LCLU map via manual/visual interpretation of the DOQQs at the stratified random point locations. This analysis enabled the generation of an error matrix from which we computed statistics, including the overall accuracy [72, 73]. The overall classification accuracy of the post-classification edited LCLU map was 76.5%.

## 3. Results and discussion

### 3.1 Stationary measurements

Stationary measurements were commenced past the semester-long student project to evaluate whether there was a seasonality to the UHI in BCS. As shown in **Figure 3**, no clear seasonal variation was observed. However, the five highest UHI intensities, all above 4°C, were all measured during spring-time (March and April). They occurred on clear or mostly clear-sky days after recent cold front passages, and possible reasons for high values under those conditions are discussed below. Lowest values, down to −1.3°C, indicative of a local urban cool island, occurred dominantly during morning measurements. Comparing sites and measurement times, the median morning UHI intensity was slightly, but statistically significantly lower than the median night/evening UHI intensity. Site differences were statistically insignificant.

**Figure 4** shows that the UHI intensity was slightly dependent on large-scale wind directions. Northerly winds, as occur behind cold front passages, were responsible for the spring-time maxima in UHI intensity at the Trails site, but no such significant difference could be found for the Quad site. This could be due to the fact that a mid-size mall, a 0.4 $km^2$ large area of impervious surface area, lies just beyond a narrow green corridor north of the Trails site, but no such prominent heat source lies near the Quad site.

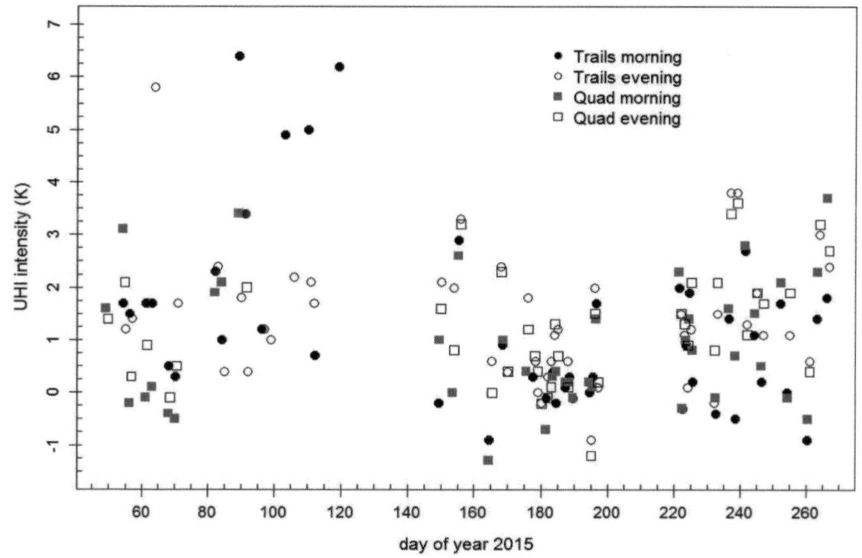


**Figure 3.**
*Seasonal changes of UHI intensity for the stationary sites and measurement times in 2015.*

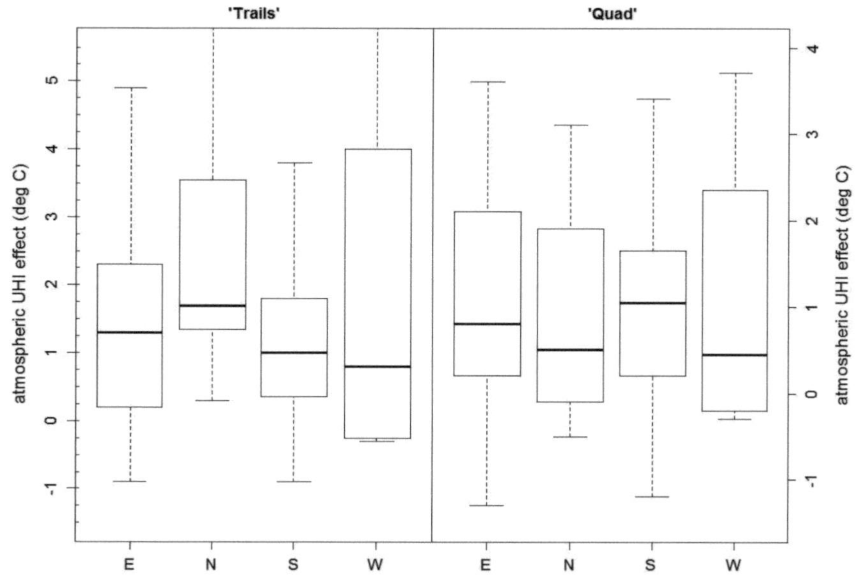


**Figure 4.**
*Boxplot of all 2015 UHI intensity measurements as a function of cardinal wind direction during the observation period. Thick horizontal bars mark medians, box edges mark the interquartile range (IQR), and whiskers are 95% confidence interval estimates.*

We used a multi-linear regression to evaluate whether and which meteorological factors may have contributed to the UHI intensity at each site. Eliminating wind direction, we found that lower temperature, and lower relative humidity days were significantly ($p > 0.95$) correlated with the UHI intensity, a result strongly driven by the high spring values though. In addition, days with higher pressures were significantly correlated with the observed UHI intensity at the Quad site ($p > 0.95$). However, no statistically significant relationship with (rural) wind speed, ranging from 0 to 7 m $s^{-1}$, was found. In summary, no single meteorological parameter other than wind direction (**Figure 4**) stood out in explaining the observations.

The observed UHI intensity range is comparable with similar measurements in other mid-size urban areas (e.g. [33, 39, 74]). However, parameters other than population (or city) size and meteorological conditions are typically more relevant, particularly impervious area fraction, anthropogenic heat release, and urban morphology [29]. Therefore, we can surmise that because both our sites have a higher than average canyon aspect ratio (building height, $H$, to canyon width, $W$, ratio), and higher than average population and thus energy use density, morning cool island findings could be related to shading and strong heat admittance into the urban fabric [75–77], while the highest UHI intensities may have been related to high local anthropogenic heat emissions, respectively. This will be discussed in more detail with respect to LCZ classifications elsewhere.

### 3.2 Mobile measurements

The selected driving routes covered central areas north and south of the city border between Bryan and College Station, where impervious surface areas maximize (LCZ 3 and 5). They also covered residential areas in south College Station, where LCZ 6 dominates.

**Figure 5** shows an overview of results from three representative late evening drives, depicting three of the four routes. These drives occurred on (from north to south) September 16, August 28, and October 14, 2015. Rural wind directions and speeds for those drives were SE at 3–4 m $s^{-1}$, ESE at 1–2 m $s^{-1}$, and ESE at 3–4 m $s^{-1}$, respectively.

In total, 15 traverses were completed at night and 16 during daytime (12 afternoon and 4 mornings). The nighttime traverses showed UHI intensities of 3.2 ± 1.9 K (mean + − 1 sd), while the daytime UHI intensities were lower at

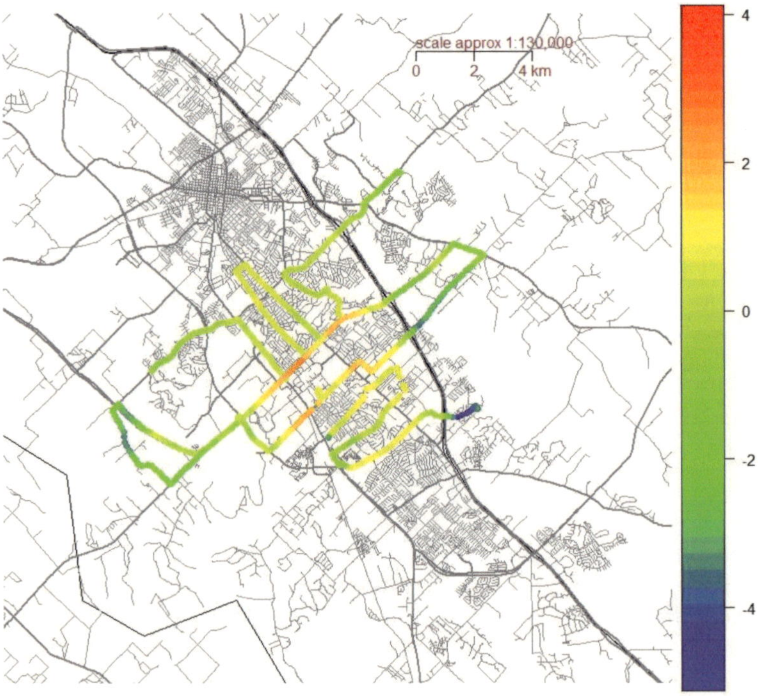


**Figure 5.**
*Roadmap based view of the BCS urban metro area overlaid with temperature anomalies (in Kelvin) during three representative late evening traverses. All traverses occurred under weak southeasterly winds. The central traverse on August 26, 2015 had the lowest wind speeds and a UHI intensity of 6.1 K.*

**Figure 6.**
*Sectional GE overlay of mobile air temperature measurements during the late evening traverse on August 26, 2015. The white arrow shows wind speed and direction during the traverse. Driving direction is indicated by the light blue arrows, and relative driving speed can be deduced from symbol spacing. Symbols designate temperatures ranging from 29.5° C (red) in the road sections downwind densely build-up urban areas 1–2 km to the west of highway 6, to 26° C (yellow) downwind the irrigated outdoor sports complex near the center of the map, to 24° C (green) downwind of the then (2015) undeveloped area southeast of the sports complex.*

2.3 ± 1.2 K. Only the nighttime UHI intensities displayed clear relationships to the prevailing meteorological conditions, namely atmospheric pressure and wind speed ($p > 0.99$), but also air temperature itself ($p > 0.95$). In addition, a weak dependence on the day's solar radiation totals was observed ($p > 0.77$). The daytime traverses' UHI intensities were not significantly related to the meteorological data; only air temperature ($p > 0.87$) and solar radiation ($p > 0.78$) were weakly related to UHI intensity for the afternoon traverses.

The combined data maintained dependencies on wind speed ($p > 0.95$) and atmospheric pressure ($p > 0.9$), and, together with the above, these results indicate that the UHI intensity of this midsize urban area was significantly more pronounced at night- as compared with daytime conditions. The day-night difference, as well as the relationships with wind speed, atmospheric pressure, and related cloud cover, reproduced findings from previous UHI studies [1].

The results from the mobile measurements will be used further to analyze the major urban drivers of UHI intensity in terms of LCZ type [8, 52, 57, 63, 78–84], and associated impervious area fractions and aspect ratios [29, 85–87].

Here, we show two case studies: one using a typical Google Earth (GE) view, and one using remotely sensed urban land cover. **Figure 6** shows a GE view of a part of the eastern BCS urban area, where the local highway separated a densely build-up area from a large outdoor sports facility, featuring frequent lawn irrigation and open, mostly undeveloped land to its southeast.

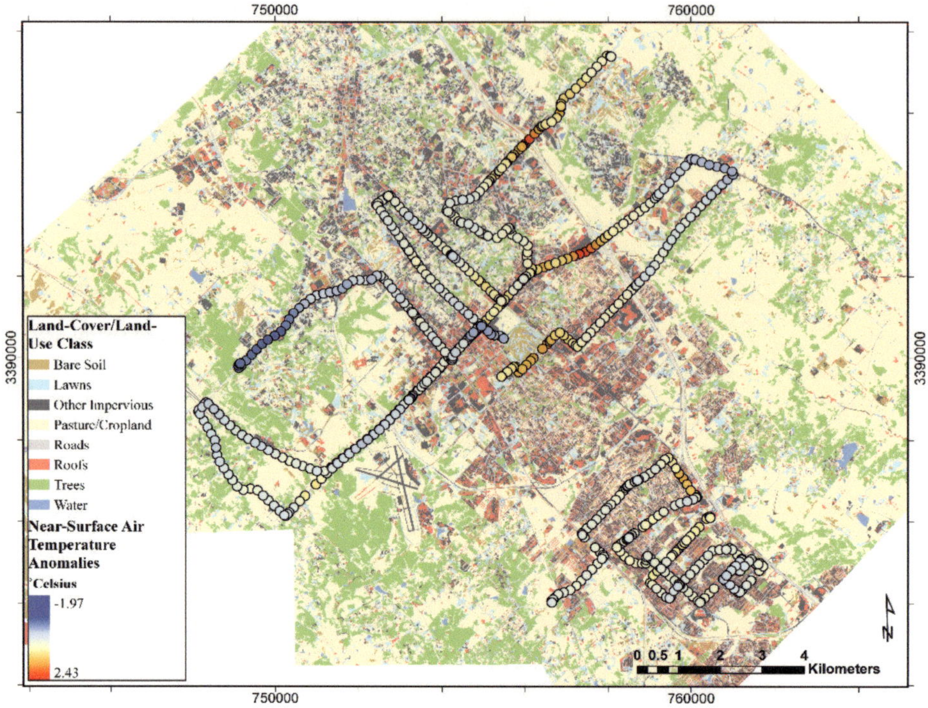


**Figure 7.**
*Temperature anomalies from several daytime (afternoon) traverses, overlaid on the LCLU map of the BCS area. The depicted routes were driven on September 2, 2015 (central SW–NE extension; winds E at 4 m $s^{-1}$), October 14, 2015 (southern extension; winds SE at 4 m $s^{-1}$), and August 26, 2015 (northern extension; winds S at 3.5 m $s^{-1}$). X and y-axis units are in meters and reflect* easting *and* northing *of UTM zone R14 (i.e., distance, respectively, from the central meridian of the zone and from the equator).*

Temperatures dropped 3°C crossing the highway toward the east into a less developed area, and another 2°C once downwind of a largely undeveloped area at the time (LCZ B/C/D). Driving direction and symbol spacing highlighted in this close-up illustrate the delayed response of the temperature sensor at higher driving speeds on the southern leg as compared with the northern leg, which included two short stops at traffic lights, one before crossing the highway, and another north of the sports complex near the center of the map.

Case study 2 is shown in **Figure 7**, which depicts temperature anomalies from three independent daytime traverses overlaid on the high resolution LCLU map. The daytime traverses showed a much more sophisticated, smaller scale pattern of the UHI as compared with the nighttime traverses. Wind speed during all these drives was south to southeasterly, and upon closer inspection the data revealed impacts of impervious surfaces areas near and upwind of the driving location [38, 88] as an important factor in determining UHI intensity in the BCS metro area. Research is ongoing to improve the LCLU map, and quantify UHI intensity as a function of the footprint's impervious area fraction, as well as other LCZ properties such as canyon aspect ratio and surface albedo.

## 4. Conclusions

We have translated an undergraduate course project of stationary measurements of the urban heat island in a mid-size urban area in east Texas into a more detailed undergraduate study of the cities' UHI intensity as a function of meteorological conditions and urban morphology. Our study was able to reproduce the expected UHI intensity relationships with wind speed and atmospheric pressure, emphasizing synoptic high pressure conditions with intense solar radiation as a major driver of the UHI. In addition, we may have also found effects of local anthropogenic heat release and building shading. The latter, urban morphology and its local to regional impacts, is an ongoing focus of UHI research. We find that not only on larger scales is the UHI effect advected downwind, but local temperatures during daytime are advected downwind on a sub-kilometer scale [8]. Hence, the fraction of impervious area immediately upwind may more strongly affect local air temperature than the local impervious area fraction at the point of measurement. Future work is intended to better quantify this effect for the urban area studied. If confirmed, and of significant size, this opens avenues for local area UHI effect mitigation planning. It suggests that previous calls for (i) more surface unsealing to reduce urban runoff and increase surface moisture and (ii) increased tree cover for additional shading and latent heat cooling, may indeed be the two most efficient pathways to reduce the UHI locally, at least in mid-size metropolitan areas, where LCZs 5 and 6 dominate.

## Acknowledgements

We are grateful to the College of Geosciences, Texas A&M University, for funding the two undergraduate students involved in this project.

## Author details

Kristen Koch[1], Gunnar W. Schade[2]*, Anthony M. Filippi[3,4], Garrison Goessler[3,4] and Burak Güneralp[3,4]

1 Environmental Programs, Texas A&M University, College Station, USA

2 Department of Atmospheric Sciences, Texas A&M University, College Station, USA

3 Department of Geography, Texas A&M University, College Station, USA

4 Center for Geospatial Sciences, Applications and Technology (GEOSAT), Texas A&M University, College Station, USA

*Address all correspondence to: gws@geos.tamu.edu

## References

[1] Arnfield AJ. Two decades of urban climate research: A review of turbulence, exchanges of energy and water, and the urban heat island. International Journal of Climatology. 2003;**23**(1):1-26

[2] Mohajerani A, Bakaric J, Jeffrey-Bailey T. The urban heat island effect, its causes, and mitigation, with reference to the thermal properties of asphalt concrete. Journal of Environmental Management. 2017;**197**:522-538

[3] Akbari H, Kolokotsa D. Three decades of urban heat islands and mitigation technologies research. Energy and Buildings. 2016;**133**:834-842

[4] Tzavali A, Paravantis JP, Mihalakakou G, Fotiadi A, Stigka E. Urban heat island intensity: A literature review. Fresenius Environmental Bulletin. 2015;**24**(12B):4535-4554

[5] Lee JS, Kim JT, Lee MG. Mitigation of urban heat island effect and greenroofs. Indoor and Built Environment. 2014;**23**(1):62-69

[6] Gago EJ, Roldan J, Pacheco-Torres R, Ordonez J. The city and urban heat islands: A review of strategies to mitigate adverse effects. Renewable & Sustainable Energy Reviews. 2013;**25**:749-758

[7] Stewart ID. A systematic review and scientific critique of methodology in modern urban heat island literature. International Journal of Climatology. 2011;**31**(2):200-217

[8] Stewart ID, Oke TR. Local climate zones for urban temperature studies. Bulletin of the American Meteorological Society. 2012;**93**(12):1879-1900

[9] Zhou DC, Zhao SQ, Liu SG, Zhang LX, Zhu C. Surface urban heat island in China's 32 major cities: Spatial patterns and drivers. Remote Sensing of Environment. 2014;**152**:51-61

[10] Kakoniti A, Georgiou G, Marakkos K, Kumar P, Neophytou MKA. The role of materials selection in the urban heat island effect in dry mid-latitude climates. Environmental Fluid Mechanics. 2016;**16**(2):347-371

[11] Morini E, Touchaei AG, Castellani B, Rossi F, Cotana F. The impact of albedo increase to mitigate the urban Heat Island in Terni (Italy) using the WRF model. Sustainability. 2016;**8**(10):999

[12] Wang YP, Akbari H. Analysis of urban heat island phenomenon and mitigation solutions evaluation for Montreal. Sustainable Cities and Society. 2016;**26**:438-446

[13] Arnfield AJ, Grimmond CSB. An urban canyon energy budget model and its application to urban storage heat flux modeling. Energy and Buildings. 1998;**27**(1):61-68

[14] Salamanca F, Martilli A, Yague C. A numerical study of the urban Heat Island over Madrid during the DESIREX (2008) campaign with WRF and an evaluation of simple mitigation strategies. International Journal of Climatology. 2012;**32**(15):2372-2386

[15] Ichinose T, Shimodozono K, Hanaki K. Impact of anthropogenic heat on urban climate in Tokyo. Atmospheric Environment. 1999;**33**(24-25):3897-3909

[16] Hinkel KM, Nelson FE, Klene AF, Bell JH. The urban heat island in winter at Barrow, Alaska. International Journal of Climatology. 2003;**23**(15):1889-1905

[17] Kim Y-H, Baik J-J. Spatial and temporal structure of the urban Heat

Island in Seoul. Journal of Applied Meteorology. 2005;**44**(5):591-605

[18] Hinkel KM, Nelson FE. Anthropogenic heat island at Barrow, Alaska, during winter: 2001-2005. Journal of Geophysical Research-Atmospheres. 2007;**112**(D06118). DOI: 10.1029/2006JD007837

[19] Lee SH, McKeen SA, Sailor DJ. A regression approach for estimation of anthropogenic heat flux based on a bottom-up air pollutant emission database. Atmospheric Environment. 2014;**95**:629-633

[20] Boehme P, Berger M, Massier T. Estimating the building based energy consumption as an anthropogenic contribution to urban heat islands. Sustainable Cities and Society. 2015;**19**:373-384

[21] Chapman S, Watson JEM, McAlpine CA. Large seasonal and diurnal anthropogenic heat flux across four Australian cities. Journal of Southern Hemisphere Earth Systems Science. 2016;**66**(3):342-360

[22] Zhou DC, Zhang LX, Hao L, et al. Spatiotemporal trends of urban heat island effect along the urban development intensity gradient in China. Science of the Total Environment. 2016;**544**:617-626

[23] Park C, Schade GW, Werner ND, Sailor DJ, Kim CH. Comparative estimates of anthropogenic heat emission in relation to surface energy balance of a subtropical urban neighborhood. Atmospheric Environment. 2016;**126**:182-191

[24] Vesala T, Jarvi L, Launiainen S, et al. Surface-atmosphere interactions over complex urban terrain in Helsinki, Finland. Tellus B. 2008;**60**(2):188-199

[25] Jarvi L, Grimmond CSB, Taka M, et al. Development of the surface urban energy and water balance scheme (SUEWS) for cold climate cities. Geoscientific Model Development. 2014;**7**(4):1691-1711

[26] Ao XY, Grimmond CSB, Chang YY, et al. Heat, water and carbon exchanges in the tall megacity of Shanghai: Challenges and results. International Journal of Climatology. 2016;**36**(14):4608-4624

[27] Loridan T, Lindberg F, Jorba O, et al. High resolution simulation of the variability of surface energy balance fluxes across Central London with urban zones for energy partitioning. Boundary-Layer Meteorology. 2013;**147**(3):493-523

[28] Ward HC, Kotthaus S, Jarvi L, Grimmond CSB. Surface urban energy and water balance scheme (SUEWS): Development and evaluation at two UK sites. Urban Climate. 2016;**18**:1-32

[29] Ryu YH, Baik JJ. Quantitative analysis of factors contributing to urban Heat Island intensity. Journal of Applied Meteorology and Climatology. 2012;**51**(5):842-854

[30] Holmer B, Eliasson I. Urban-rural vapour pressure differences and their role in the development of urban heat islands. International Journal of Climatology. 1999;**19**(9):989-1009

[31] Runnalls KE, Oke TR. Dynamics and controls of the near-surface heat island of Vancouver, British Columbia. Physical Geography. 2000;**21**(4):283-304

[32] Goldreich Y. Urban climate studies in Johannesburg, a Suptropical City located on a ridge - a review. Atmospheric Environment Part B-Urban Atmosphere. 1992;**26**(3):407-420

[33] Park H-S. Features of the heat island in Seoul and its surrounding cities. Atmospheric Environment (1967). 1986;**20**(10):1859-1866

[34] Saitoh TS, Shimada T, Hoshi H. Modeling and simulation of the Tokyo urban heat island. Atmospheric Environment. 1996;**30**(20):3431-3442

[35] Akinbode OM, Eludoyin AO, Fashae OA. Temperature and relative humidity distributions in a medium-size administrative town in Southwest Nigeria. Journal of Environmental Management. 2008;**87**(1):95-105

[36] Tso CP. A survey of urban heat island studies in two tropical cities. Atmospheric Environment. 1996;**30**(3):507-519

[37] Deosthali V. Impact of rapid urban growth on heat and moisture islands in Pune City, India. Atmospheric Environment. 2000;**34**(17):2745-2754

[38] Comrie AC. Mapping a wind-modified urban heat island in Tucson, Arizona (with comments on integrating research and undergraduate learning). Bulletin of the American Meteorological Society. 2000;**81**(10):2417-2431

[39] Yamashita S, Sekine K, Shoda M, Yamashita K, Hara Y. On relationships between heat island and sky view factor in the cities of Tama River basin, Japan. Atmospheric Environment (1967). 1986;**20**(4):681-686

[40] Honjo T, Yamato H, Mikami T, Grimmond CSB. Network optimization for enhanced resilience of urban heat island measurements. Sustainable Cities and Society. 2015;**19**:319-330

[41] Kuttler W, Barlag AB, Rossmann F. Study of the thermal structure of a town in a narrow valley. Atmospheric Environment. 1996;**30**(3):365-378

[42] Spronken-Smith RA, Oke TR. The thermal regime of urban parks in two cities with different summer climates. International Journal of Remote Sensing. 1998;**19**(11):2085-2104

[43] Upmanis H, Eliasson I, Lindqvist S. The influence of green areas on nocturnal temperatures in a high latitude city (Goteborg, Sweden). International Journal of Climatology. 1998;**18**(6):681-700

[44] Torok SJ, Morris CJG, Skinner C, Plummer N. Urban heat island features of southeast Australian towns. Australian Meteorological Magazine. 2001;**50**(1):1-13

[45] Unger J, Sumeghy Z, Zoboki J. Temperature cross-section features in an urban area. Atmospheric Research. 2001;**58**(2):117-127

[46] Jonsson P. Vegetation as an urban climate control in the subtropical city of Gaborone, Botswana. International Journal of Climatology. 2004;**24**(10):1307-1322

[47] Huang LM, Zhao DH, Wang JZ, Zhu JY, Li JL. Scale impacts of land cover and vegetation corridors on urban thermal behavior in Nanjing, China. Theoretical and Applied Climatology. 2008;**94**(3-4):241-257

[48] Sun CY, Brazel AJ, Chow WTL, Hedquist BC, Prashad L. Desert heat island study in winter by mobile transect and remote sensing techniques. Theoretical and Applied Climatology. 2009;**98**(3-4):323-335

[49] Yokobori T, Ohta S. Effect of land cover on air temperatures involved in the development of an intra-urban heat island. Climate Research. 2009;**39**(1):61-73

[50] da Silva VDR, de Azevedo PV, Brito RS, Campos J. Evaluating the urban climate of a typically tropical city of northeastern Brazil. Environmental Monitoring and Assessment. 2010;**161**(1-4):45-59

[51] Charabi Y, Bakhit A. Assessment of the canopy urban heat island of a

coastal arid tropical city: The case of Muscat, Oman. Atmospheric Research. 2011;**101**(1-2):215-227

[52] Alexander PJ, Mills G. Local climate classification and Dublin's urban Heat Island. Atmosphere. 2014;**5**(4):755-774

[53] Borbora J, Das AK. Summertime urban Heat Island study for Guwahati City, India. Sustainable Cities and Society. 2014;**11**:61-66

[54] Busato F, Lazzarin RM, Noro M. Three years of study of the urban Heat Island in Padua: Experimental results. Sustainable Cities and Society. 2014;**10**:251-258

[55] Szegedi S, Toth T, Lazar I. Role of urban morphology in development of the thermal excess in the City of Debrecen, Hungary. Environmental Engineering and Management Journal. 2014;**13**(11):2805-2808

[56] Dobrovolny P, Krahula L. The spatial variability of air temperature and nocturnal urban heat island intensity in the city of Brno, Czech Republic. Moravian Geographical Reports. 2015;**23**(3):8-16

[57] Leconte F, Bouyer J, Claverie R, Petrissans M. Using local climate zone scheme for UHI assessment: Evaluation of the method using mobile measurements. Building and Environment. 2015;**83**:39-49

[58] Czubaszek R, Wysocka-Czubaszek A. Urban Heat Island in Bialystok. Journal of Ecological Engineering. 2016;**17**(3):60-65

[59] Qaid A, Bin Lamit H, Ossen DR, Shahminan RNR. Urban heat island and thermal comfort conditions at micro-climate scale in a tropical planned city. Energy and Buildings. 2016;**133**:577-595

[60] Chakraborty T, Sarangi C, Tripathi SN. Understanding Diurnality and inter-seasonality of a sub-tropical urban Heat Island. Boundary-Layer Meteorology. 2017;**163**(2):287-309

[61] Liu L, Lin YY, Liu J, et al. Analysis of local-scale urban heat island characteristics using an integrated method of mobile measurement and GIS-based spatial interpolation. Building and Environment. 2017;**117**:191-207

[62] Dihkan M, Karsli F, Guneroglu N, Guneroglu A. Evaluation of urban heat island effect in Turkey. Arabian Journal of Geosciences. 2018;**11**(8):186. Available from: https://doi.org/10.1007/s12517-018-3533-3

[63] Shi Y, Lau KKL, Ren C, Ng E. Evaluating the local climate zone classification in high-density heterogeneous urban environment using mobile measurement. Urban Climate. 2018;**25**:167-186

[64] Yadav N, Sharma C. Spatial variations of intra-city urban heat island in megacity Delhi. Sustainable Cities and Society. 2018;**37**:298-306

[65] Ziter CD, Pedersen EJ, Kucharik CJ, Turner MG. Scale-dependent interactions between tree canopy cover and impervious surfaces reduce daytime urban heat during summer. Proceedings of the National Academy of Sciences. 2019;**116**(15):7575-7580

[66] Benz UC, Hofmann P, Willhauck G, Lingenfelder I, Heynen M. Multi-resolution, object-oriented fuzzy analysis of remote sensing data for GIS-ready information. ISPRS Journal of Photogrammetry and Remote Sensing. 2004;**58**(3-4):239-258

[67] Blaschke T. Object based image analysis for remote sensing. ISPRS Journal of Photogrammetry and Remote Sensing. 2010;**65**(1):2-16

[68] Rouse JW, Haas RH, Schell JA, Deering DW, Freden SC, Mercanti EP,

et al., editors. Monitoring Vegetation Systems in the Great Plains with ERTS. Greenbelt: NASA; 1974

[69] McFeeters SK. The use of the normalized difference water index (NDWI) in the delineation of open water features. International Journal of Remote Sensing. 1996;**17**(7):1425-1432

[70] Congalton RG. A review of assessing the accuracy of classifications of remotely sensed data. Remote Sensing of Environment. 1991;**37**(1):35-46

[71] Congalton RG, Green K. Assessing the Accuracy of Remotely Sensed Data: Principles and Practices. Bocan Raton: Lewis Publishers; 1999

[72] Filippi AM, Jensen JR. Fuzzy learning vector quantization for hyperspectral coastal vegetation classification. Remote Sensing of Environment. 2006;**100**(4):512-530

[73] Filippi AM, Jensen JR. Effect of continuum removal on hyperspectral coastal vegetation classification using a fuzzy learning vector quantizer. IEEE Transactions on Geoscience and Remote Sensing. 2007;**45**(6):1857-1869

[74] Oke TR. City size and the urban heat island. Atmospheric Environment (1967). 1973;**7**(8):769-779

[75] Santamouris M. Analyzing the heat island magnitude and characteristics in one hundred Asian and Australian cities and regions. Science of the Total Environment. 2015;**512**:582-598

[76] Alonso MS, Fidalgo MR, Labajo JL. The urban heat island in Salamanca (Spain) and its relationship to meteorological parameters. Climate Research. 2007;**34**(1):39-46

[77] Giovannini L, Zardi D, de Franceschi M. Analysis of the urban thermal fingerprint of the city of Trento in the Alps. Journal of Applied Meteorology and Climatology. 2011;**50**(5):1145-1162

[78] Cai M, Ren C, Xu Y, Dai W, Wang XM. In: Lai PC, Low CT, Wong PPY, editors. International Conference on Geographies of Health and Living in Cities: Making Cities Healthy for all. 2016;**36**(iii-iv):82-89

[79] Cai M, Ren C, Xu Y. 2017 Joint Urban Remote Sensing Event. IEEE; 2017

[80] Cardoso RD, Amorim M. Urban heat island analysis using the 'local climate zone' scheme in Presidente Prudente, Brazil. Investigaciones Geograficas-Spain. 2018;**69**:107-118

[81] Lin ZL, Xu HQ. In: Weng Q, Gamba P, Xian G, Chen JM, Liang S, editors. A study of urban heat island intensity based on local climate zones: A case study in fuzhou, china. 4th International Workshop on Earth Observation and Remote Sensing Applications. Guangzhou, China; 2016:250-254. DOI: 10.1109/EORSA.2016.7552807

[82] Skarbit N, Stewart ID, Unger J, Gal T. Employing an urban meteorological network to monitor air temperature conditions in the 'local climate zones' of Szeged, Hungary. International Journal of Climatology. 2017;**37**:582-596

[83] Wang CY, Middel A, Myint SW, et al. Assessing local climate zones in arid cities: The case of Phoenix, Arizona and Las Vegas, Nevada. ISPRS Journal of Photogrammetry and Remote Sensing. 2018;**141**:59-71

[84] Wang ZH, Xing W, Huang Y, Xie TA. Studying the urban Heat Island using a local climate zone scheme. Polish Journal of Environmental Studies. 2016;**25**(6):2609-2616

[85] Kolokotroni M, Giridharan R. Urban heat island intensity in London: An

investigation of the impact of physical characteristics on changes in outdoor air temperature during summer. Solar Energy. 2008;**82**(11):986-998

[86] Memon RA, Leung DYC, Liu CH. Effects of building aspect ratio and wind speed on air temperatures in urban-like street canyons. Building and Environment. 2010;**45**(1):176-188

[87] Theeuwes NE, Steeneveld GJ, Ronda RJ, et al. Seasonal dependence of the urban heat island on the street canyon aspect ratio. Quarterly Journal of the Royal Meteorological Society. 2014;**140**(684):2197-2210

[88] Heaviside C, Cai XM, Vardoulakis S. The effects of horizontal advection on the urban heat island in Birmingham and the west midlands, United Kingdom during a heatwave. Quarterly Journal of the Royal Meteorological Society. 2015;**141**(689):1429-1441

Chapter 3

# Mapping and Estimation of Nitrogen and Sulfur Atmospheric Deposition Fluxes in Central Region of the Mexican Bajio

*Rosa María Cerón Bretón, Julia Griselda Cerón Bretón, Reyna del Carmen Lara Severino, Marcela Rangel Marrón, María de la Luz Espinosa Fuentes, Simón Eduardo Carranco Lozada and Lizbeth Cisneros Rosique*

## Abstract

The objective of this study was to assess the spatial and temporal distribution of nitrogen and sulfur deposition and its relationship with meteorological conditions in the metropolitan area of León in Guanajuato, México. N and S atmospheric deposition was collected using passive samplers (IER) in 10 sites in León City during three climatic seasons in 2018. Nitrate, ammonium, and sulfate concentrations and deposition fluxes of N and S were determined. From wind and air-mass trajectories analysis, mechanisms and possible sources contributing to N and S deposition in the study area were assessed. Atmospheric deposition fluxes were compared to critical load values reported for sensitive ecosystems in Europe. It was found that mean deposition flux for N (5.82 Kg N $ha^{-1}$ $year^{-1}$) was within the range of values reported for sensitive ecosystems in Nuevo México, Europe, and California. On the other hand, the mean deposition flux for S (13.77 S Kg $ha^{-1}$ $year^{-1}$) exceeded the critical load values proposed for Europe, suggesting that current S deposition could be a risk for ecosystems and water bodies in the region.

**Keywords:** atmospheric deposition, nitrogen, sulfur, Guanajuato, México

## 1. Introduction

Air pollution constitutes an important health risk not only for developed countries but also for developing countries. $SO_2$ and $NO_2$ are among primary pollutants commonly present in urban and industrial areas, whose main sources are combustion processes, and being both, acid rain precursors. The two main mechanisms responsible for pollutant removal in the atmosphere are dry and



IntechOpen

wet deposition. Gases and particulate matter are deposited in dry form, while completely soluble components are deposited by the rain action (wet deposition). Dry deposition of gases and particles occurs through complex processes such as sedimentation, impactation, and adsorption and can contribute in a significant way to local deposition of atmospheric components, being particularly important near urban and industrial areas, where particle concentration and dust can be relatively high [1]. On the other hand, wet deposition occurs as rain, fog or snow and plays an important role in the removal processes of soluble wastes in the atmosphere [2]. Some of components of atmospheric deposit (as nitrate and sulfate) are precursors of acidification, being able to harm the aquatic and terrestrial ecosystems not only in the surroundings of sources but also in nearby regions.

Thus, it is necessary to measure the total atmospheric deposition of N and S, to estimate the inflows and outflows, their effects on the balance of biogeochemical cycles, and to assess the biologic and ecologic response to the current atmospheric pollution levels and their relationship with emission patterns. Since the effective design of public politics requires of surveillance and monitoring programs to understand and to quantify the current conditions, by comparison with historical data and reference values, measurement chemical composition, and physical characteristics are required. This will allow to propose environmental politics focused to protect not only public health but also ecosystems and diagnose the real effects of N and S deposition as a result of the current emission patterns [3]. Deposition maps are very useful tools to carry out this kind of assessments, since they allow to analyze the spatial distribution and temporal variability of N and S deposition in a given area; likewise, they allow to visualize those areas in which exceedances to threshold values of critical loads are occurring. In this way, the decision makers can implement control strategies of regionalized emissions to protect different receptors. Several exceedance ranges can be established and be related to sensitivity categories, from which, it is possible to diagnose in a preliminary way, the vulnerability of ecosystems to the inputs of N and S. However, the main problem during this process is to have enough and reliable data of N and S deposition fluxes.

Dry deposition fluxes for ecosystems are obtained from theoretical models, due in part to the lack of monitoring standardized methods. On the other hand, in the case of wet deposition, the estimation depends on the occurrence of rain events, resulting in the case of arid or semiarid regions in insufficient data to study the spatial and temporal variability at long-term. In addition, atmospheric sampling devices for atmospheric deposition comprise manual and automatic collectors. Although the first choice in more economical, it is hard to implement in field in the case of remote sites, while in the case of automatic collectors, these are expensive and must meet specific criteria for installation and operation. Fenn and Bytnerowicz [4] proposed a collector based on ionic exchange resin (IER) to quantify atmospheric deposition in forests. In this regard, comparison between conventional collectors and passive sampling devices of atmospheric deposition have been carried out [5, 6], concluding that passive collectors can be used to quantify total sulfur deposition and report a strong correlation between N Deposition and the presence of nitrate in soils. From combined use of mapping and measurement of N and S deposition fluxes, it is possible to study the spatial and temporal variability of deposition. Therefore, this study was aimed to estimate N and S atmospheric deposition fluxes using passive collectors to assess their spatial distribution and temporal variability in Central Region of México known as "Bajío."

## 2. Materials and methods

### 2.1 Study area

Study area is located in Central México within an area known as "Bajío," in metropolitan area of León, in Guanajuato. León City is the most populated in Guanajuato state with a population of 1,578,626 inhabitants [7]. Climate is semi-dry semi-warm [8].

The main economic activities in this region are shoe manufacturing, tanned skin, cardboard production, and chemical industry; however, economical activities are not only based on these sectors, but also building, plastics, mining, manufacturing, textile, and automotive industries. There were selected a total of 10 sampling sites considering in first place the site accessibility and the land use; as well as safety of sampling devices, preferring public buildings to safely house the equipment. In **Figure 1**, study area and sampling site location are shown.

### 2.2 Sampling

Since passive sampling is defined as hydrologic flux to soil of ions and other compounds in solution, passive sampling provides a useful estimation of atmospheric inputs to a given site, because of it includes both, wet and dry deposition. Considering the high cost and the difficulty of measuring dry deposition fluxes, passive collectors constitute a useful alternative to measure annual atmospheric inputs at ground level [4, 9]. Additionally, automatic collectors are very expensive, for this reason, passive collectors constitute a good sampling choice in a given area, since they allow to increase the number of sampling sites at a low cost and take samples simultaneously in different locations in a specific region.

Therefore, it is possible to obtain complex spatial patterns of N and S atmospheric deposition in a given area by using monitoring equipment of low cost, easy to operate, and that does not require frequent field visits. Collectors based on IER

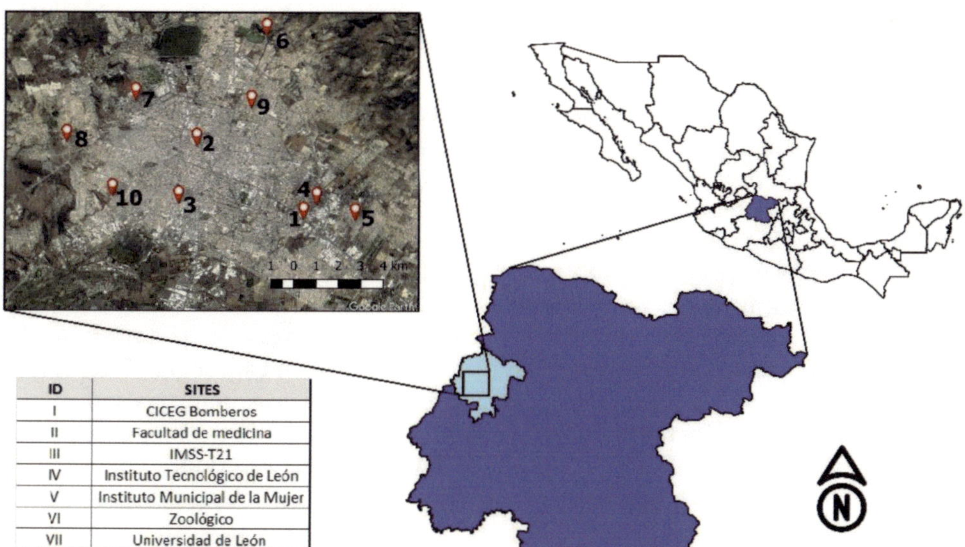


| ID | SITES |
|---|---|
| I | CICEG Bomberos |
| II | Facultad de medicina |
| III | IMSS-T21 |
| IV | Instituto Tecnológico de León |
| V | Instituto Municipal de la Mujer |
| VI | Zoológico |
| VII | Universidad de León |
| VIII | CECYT |
| IX | La alameda |
| X | Loma Blanca |

**Figure 1.**
*Study area location.*

beds have been used to measure atmospheric deposition in forest ecosystems with a high spatial resolution [10].

This type of passive collectors consists in a funnel connected to a column which contains 30 g of ionic exchange resin (IER) [11]. Glass wool is placed in both, bottom (as a support) and top (as a filter) sections. Samples are collected through the funnel, and hydrologic flux is channeled to resin mixed bed through the column where ions are retained. The funnel is connected to IER column by a PVC tube (1.27 cm × 35.6 cm), in addition, a double-wall shadow tube is placed around resin column to avoid that solar radiation damage the resin. Resin used for IER collectors was a mixed bed of polystyrene to exchange both, anions and cations (Amberlite IRN-150™). In the top of sampling devices, a fine mesh is placed to avoid the input of insects, leaves, and so on. The PVC column has a valve which must be open all time to let the pass of hydrologic flux through the collector.

The main advantage of this kind of collector is that can be used during long exposition periods (i.e., months), the equipment has a low cost, and let to display a great number of them to characterize deposition spatial patterns with a high resolution [12].

Sampling was carried out from January 1 to December 31 of 2018, with three sampling sub-periods considering three climatic seasons: cold dry, warm dry, and rains. At the end of each sampling sub-period, resin tubes were changed by tubes containing fresh resin. Later, retained ions (sulfate, nitrate, and ammonium) were extracted from resin tube by using an extraction solution (KCL 2N solution) and analyzed by turbidimetric and colorimetric methods, respectively.

### 2.3 Chemical analysis

Atmospheric deposition samples were sent and analyzed in Environmental Protection Laboratory of Chemistry Faculty of Autonomous University of Carmen. $NH_4^+$ was determined by molecular absorption spectrometry by using blue indophenol method [13]. $SO_4^{2-}$ was determined by turbidimetric method [14], whereas $NO_3^-$ was analyzed by colorimetric method using the brucine method [15].

### 2.4 Meteorological analysis

To identify possible natural and anthropogenic sources and to analyze transport processes which could influence on N and S levels in atmospheric deposition in the study area, a meteorological analysis was carried out at both, surface and altitude. For this, surface meteorological data were obtained from meteorological portable stations (Davies Vantage Pro II). Later, to carry out surface meteorological analysis, wind rose graphs were obtained by using WRPLOT View™ (Lakes Environmental), to identify prevailing wind direction in the study area.

For the altitude meteorological analysis, backward air-masses trajectories for 24 h were obtained from HYSPLIT model (Hybrid Single-Particle Lagrangian Integrated Trajectory) from NOAA (National Oceanic and Atmospheric Administration of United States) to identify the origin of air masses and to identify the main transport processes contributing to N and S deposition during the study period.

### 2.5 Statistical analysis

To obtain descriptive measurements, to analyze the morphology and symmetry of data, univariate, bivariate, and multivariate analysis was carried out by using XLSTAT 20016 program. Likewise, non-parametric tests (Friedman test) were applied to establish if, there were significant differences between treatments (sites and sampling seasons) [16].

### 2.6 Kriging interpolation and mapping deposition fluxes

To obtain deposition maps in the study area, a geo-statistical procedure was applied to interpolate field measurements (kriging interpolation) and to obtain a continuous spatial pattern for the variables (concentration iso-lines) to increase the number of points in the maps.

## 3. Results and discussion

### 3.1 Sulfate

#### *3.1.1 Seasonal variability*

To assess the seasonal variability, deposition fluxes were analyzed for three climatic periods: cold dry, warm dry, and rainy. During the cold dry season, a mean flux of 10.89 Kg $Ha^{-1}$ $year^{-1}$, with a maximum of 17.14 Kg $Ha^{-1}$ $year^{-1}$ in site VI, which corresponds to Zoológico, located at NE, whereas the minimum value of 6.58 Kg $Ha^{-1}$ $year^{-1}$ was obtained for site VII that corresponds to Universidad de León (**Figure 2**).

On the other hand, the mean value obtained during warm dry season was 13.98 Kg $Ha^{-1}$ $year^{-1}$, with a maximum value of 25.82 Kg $Ha^{-1}$ $year^{-1}$ in site IV (Instituto Tecnológico de León), located at SE, whereas the minimum value (1.81 Kg $Ha^{-1}$ $year^{-1}$) was found in site III, which corresponds to monitoring station IMSS T-21, located at downtown of the city (**Figure 2**).

Finally, during rainy season, the mean value obtained was 15.52 Kg $Ha^{-1}$ $year^{-1}$ with a maximum of 18.43 Kg $Ha^{-1}$ $year^{-1}$ in sites III and VI, located at NE and at downtown of the city, whereas the minimum value (11.77 Kg $Ha^{-1}$ $year^{-1}$) was obtained for site I (monitoring station CICEG-Bomberos) (**Figure 2**).

From **Figure 2**, it can be observed that mean deposition fluxes for sulfate were relatively higher during the rainy season and lower during cold dry season. With respect to extreme values, the highest values were observed during warm dry

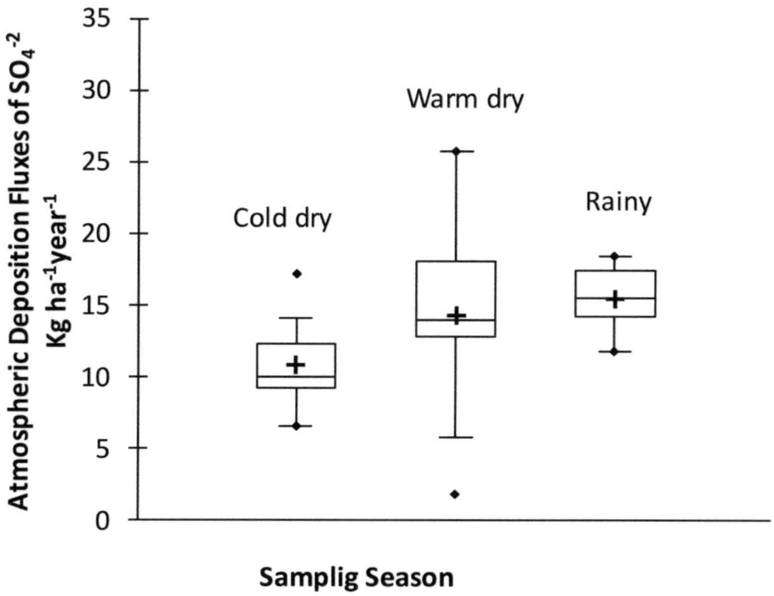


**Figure 2.**
*Sulfate atmospheric deposition fluxes by season.*

season, suggesting that the lack of dilution during this season could result in higher values in the region. Applying the Friedman tests, significant differences in sulfate atmospheric deposition fluxes were found between seasons.

### *3.1.2 Spatial variability*

To assess the spatial distribution, sulfate atmospheric deposition fluxes were analyzed by site. A mean value of 13.78 Kg $Ha^{-1}$ $year^{-1}$ was found, with a maximum value of 25.82 Kg $Ha^{-1}$ $year^{-1}$. From **Figure 3**, it can be observed that the highest mean value was found in site IV (Instituto Tecnológico de León) located at SE, whereas the minimum value (1.81 Kg $Ha^{-1}$ $year^{-1}$) was registered for site III, where monitoring station IMSS-T21 is located just at downtown of the city. Considering extreme values, the higher fluxes were found for site IV; however, when Friedman test was applied, significant differences were not found. It suggests a certain homogeneity in sulfate deposition levels in the study region, which confirms the regional character of sulfate precursor ($SO_2$).

### *3.1.3 Analysis by land use*

To assess the variability of sulfate atmospheric deposition fluxes by land use, sampling sites were grouped according to the prevailing land use as follows:

Industrial land use: Sites I, II, III, IV, V.

Urban land use: Sites VI, VII, VIII, IX, X.

According to **Figure 4**, it can be observed that mean sulfate deposition fluxes were slightly higher in sites with an urban land use; however, sites with an industrial land use showed extreme values. When Friedman tests was applied, significant differences between land use were not found. It demonstrates a spatial homogeneity that suggests the regional origin of sulfate as a result of long-range transport mainly during the rainy season.

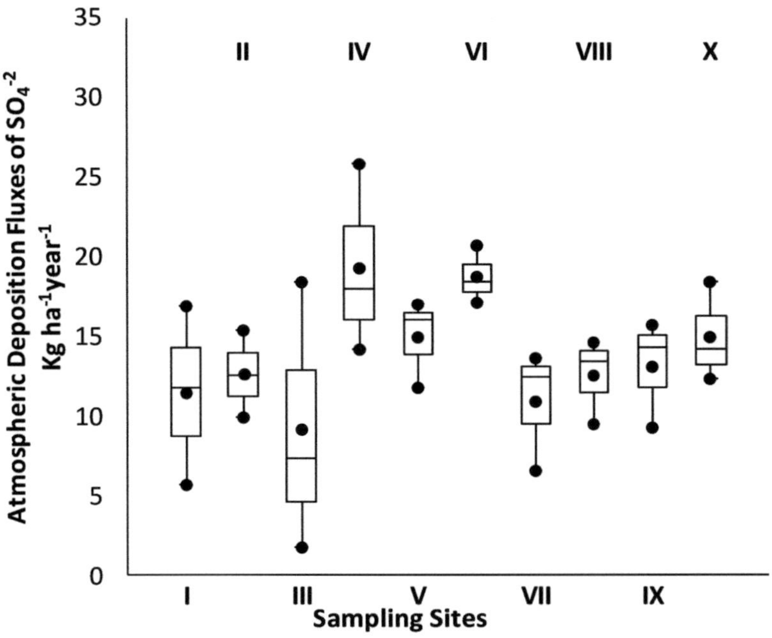


**Figure 3.**
*Sulfate atmospheric deposition fluxes by site.*

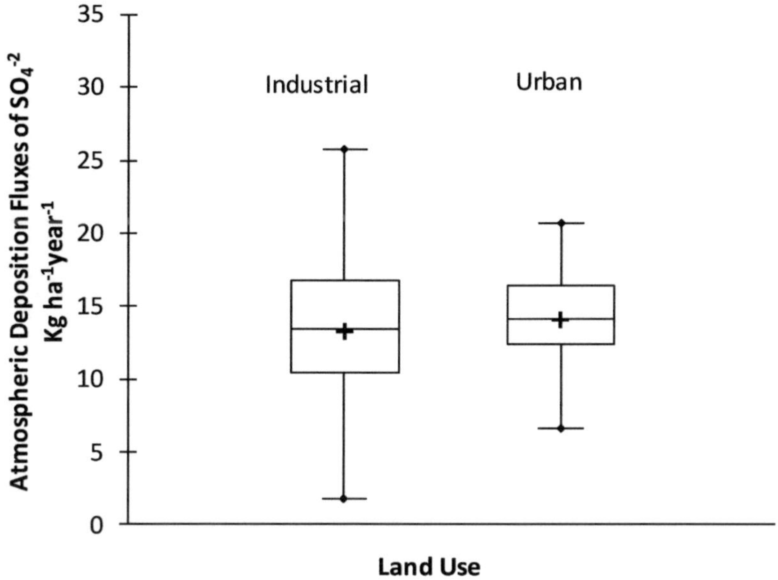


**Figure 4.**
*Sulfate atmospheric deposition fluxes by land use.*

In the case of sites grouped with an urban land use, the mean value found was 14.05 Kg $Ha^{-1}$ $year^{-1}$, with a maximum and minimum values of 20.70 and 6.58 Kg $Ha^{-1}$ $year^{-1}$, respectively, observed in sites VI (Zoológico at NE) and site VII (Universidad de León) (**Figure 4**).

On the other hand, in the case of sites grouped with an industrial land use, the mean value found was 13.51 Kg $Ha^{-1}$ $year^{-1}$, with a maximum value of 25.82 Kg $Ha^{-1}$ $year^{-1}$. Considering extreme values, the highest values were found in site IV (Instituto Tecnológico de León at SE), whereas the lowest value (1.81 Kg $Ha^{-1}$ $year^{-1}$) was registered in site III (monitoring station IMSS-T21 at downtown of the city).

## 3.2 Nitrate

### *3.2.1 Seasonal variability*

From **Figure 5**, it can be observed that the higher nitrate deposition fluxes were obtained during cold dry season, with a mean value of 2.93 Kg $Ha^{-1}$ $year^{-1}$ and a maximum value of 3.93 Kg $Ha^{-1}$ $year^{-1}$, which corresponds to site X (Loma Blanca, at SW). On the other hand, the mean value obtained for warm dry season was 0.78 Kg $Ha^{-1}$ $year^{-1}$ with a maximum of 1.47 Kg $Ha^{-1}$ $year^{-1}$ in site X (Loma Blanca at SW). Finally, during the rainy season, a mean value of 1.13 Kg $Ha^{-1}$ $year^{-1}$ with a maximum value of 2.84 Kg $Ha^{-1}$ $year^{-1}$ in site X (Loma Blanca at SW) and a minimum value of 0.22 Kg $Ha^{-1}$ $year^{-1}$ in site V (Instituto de la Mujer at SE) were found (**Figure 5**).

Applying Friedman tests, significant differences were found between cold dry season and the other two sampling seasons.

### *3.2.2 Spatial variability*

From the analysis by sampling site, a mean nitrate atmospheric deposition flux of 1.61 Kg $Ha^{-1}$ $year^{-1}$ with a maximum value of 3.93 Kg $Ha^{-1}$ $year^{-1}$ was found. The highest value was observed in site X (Loma Blanca at SW) (**Figure 6**).

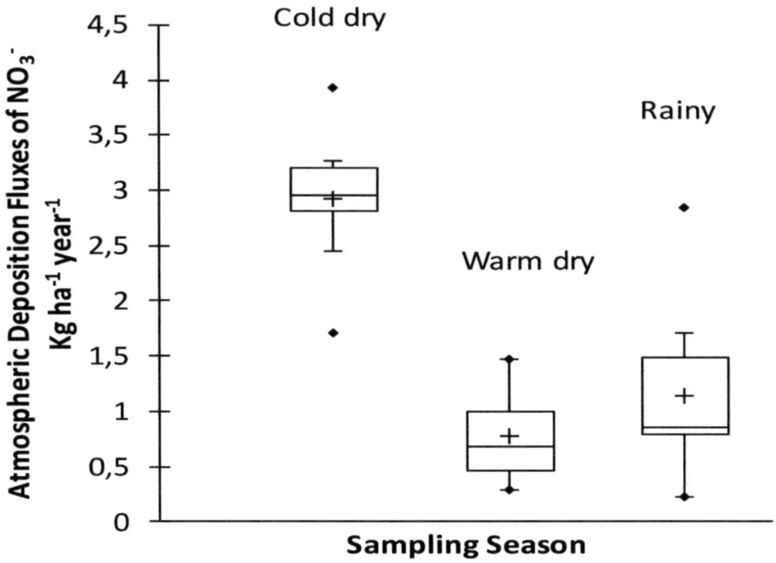


**Figure 5.**
*Nitrate atmospheric deposition fluxes by season.*

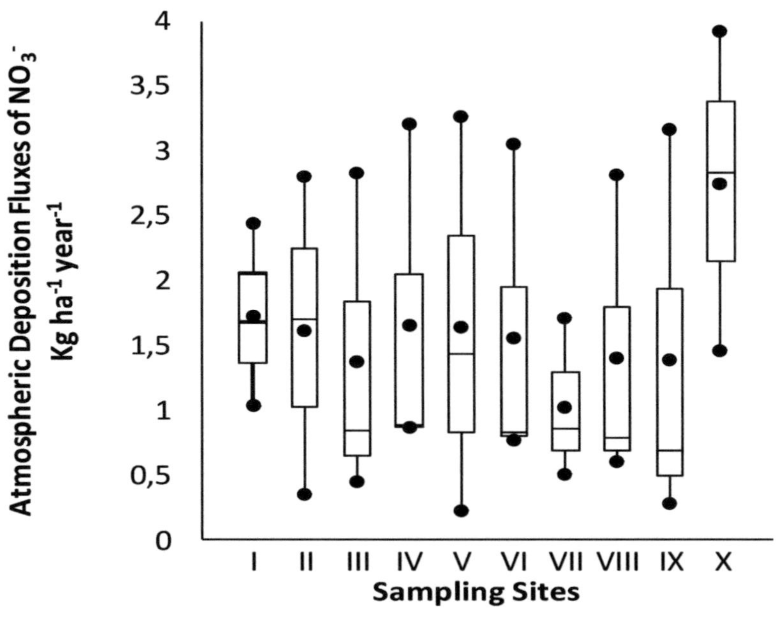


**Figure 6.**
*Nitrate atmospheric deposition fluxes by site.*

However, when Friedman test was applied, differences found were not statistically significant, suggesting a certain uniformity in the distribution of sources (especially mobile sources) and nitrate levels around study area.

### *3.2.3 Analysis by land use*

To assess the variability of nitrate atmospheric deposition fluxes by land use, sampling sites were grouped according to the prevailing land use as follows:

Industrial land use: Sites I, II, III, IV, V.

Urban land use: Sites VI, VII, VIII, IX, X.

In the case of sites grouped with an urban land use, it was obtained a mean value of 1.62 Kg $Ha^{-1}$ $year^{-1}$ with a maximum of 3.93 Kg $Ha^{-1}$ $year^{-1}$. Regarding extreme values, the highest nitrate deposition fluxes were found in site X (**Figure 7**). On the other hand, in the case of sites grouped with an industrial land use, the mean value obtained was 1.50 Kg $Ha^{-1}$ $year^{-1}$, with a maximum of 3.26 Kg $Ha^{-1}$ $year^{-1}$. The highest flux was registered in site I, corresponding to monitoring station located at CICEG Bomberos (**Figure 7**).

According to **Figure 7**, it can be observed that nitrate deposition was slightly higher in sites with an urban land use; this agrees with the origin of nitrate, whose main source is mobile sources, which are uniformly distributed along metropolitan area of León. This hypothesis was confirmed by applying Friedman tests which demonstrated that there were not significant differences between sampling sites or land use.

### 3.3 Ammonium

#### *3.3.1 Seasonal variability*

From **Figure 8**, it can be observed that the mean ammonium deposition flux was higher during the cold dry season. During the cold dry season, the mean value obtained was 11.29 Kg $Ha^{-1}$ $year^{-1}$ with a maximum value of 15.78 Kg $Ha^{-1}$ $year^{-1}$ which corresponds to sites I and IV (CICEG Bomberos and Instituto Tecnológico de León, located at SE). On the other hand, during the other two seasons, nitrate deposition fluxes were significantly lower. The mean value obtained during warm dry season was 1.08 Kg $Ha^{-1}$ $year^{-1}$ with a maximum of 6.07 Kg $Ha^{-1}$ $year^{-1}$ in site VI (Zoológico located at NE), whereas the mean value registered for rainy season was 0.25 Kg $Ha^{-1}$ $year^{-1}$ with a maximum value of 1.06 Kg $Ha^{-1}$ $year^{-1}$ in site III (IMSS-T21, at the downtown of the city) (**Figure 8**).

The marked seasonality found in ammonium levels suggests two important aspects: first, the dilution effect is important since ammonium levels during the rainy season were considerably lower as a result of frequent and intense rains

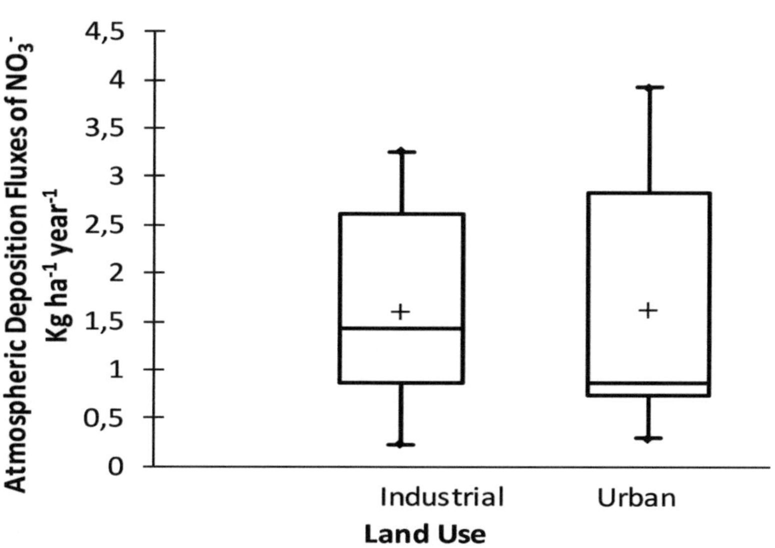


**Figure 7.**
*Nitrate atmospheric deposition fluxes by land use.*

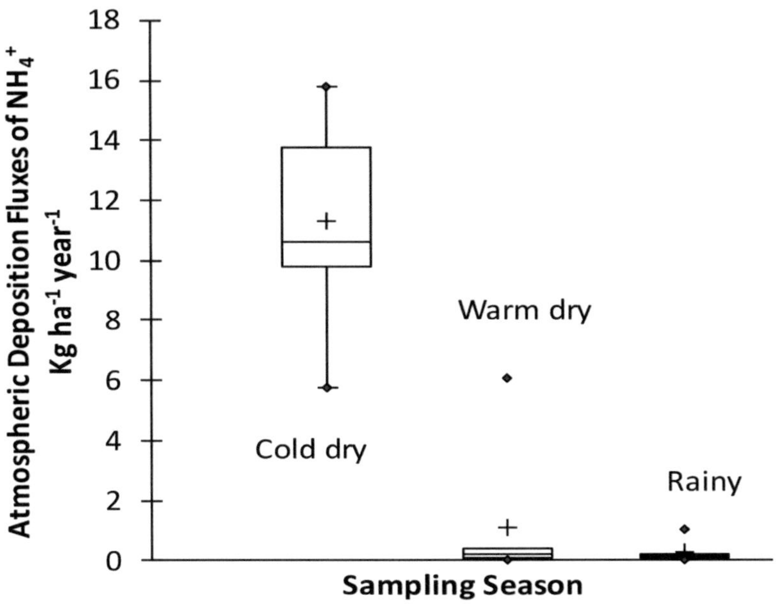


**Figure 8.**
*Ammonium atmospheric deposition fluxes by season.*

occurring during this season in comparison with the other two seasons of the year. Second, the temporality of sources could be important, since the use of agrochemicals can be intensive during the cold dry season, resulting in levels considerably high during this season. This was demonstrated applying Friedman test, whose results confirmed that differences between seasons were significant.

### *3.3.2 Spatial variability*

To assess the spatial variability and their distribution, ammonium deposition fluxes were analyzed by sampling site, finding a mean value of 4.29 Kg $Ha^{-1}$ $year^{-1}$, with the highest value registered in site X (Loma Blanca at SW) (**Figure 9**). However, applying Friedman test, it was concluded that there were not significant differences between sampling sites. It suggests that the differences found could be attributed only to sampling season as a result of agricultural activities and the effect of atmospheric dilution.

### *3.3.3 Analysis by land use*

To assess the variability of ammonium atmospheric deposition fluxes by land use, sampling sites were grouped according to the prevailing land use as follows:

Industrial land use: Sites I, II, III, IV, V.

Urban land use: Sites VI, VII, VIII, IX, X.

In the case of sites grouped with an urban land use, the mean value obtained was 3.97 Kg $Ha^{-1}$ $year^{-1}$, with a maximum of 14.04 Kg $Ha^{-1}$ $year^{-1}$. Regarding extreme values, the highest values were found in site VI (Zoológico at NE), whereas in the case of sites grouped with an industrial land use, the mean value obtained was 4.45 Kg $Ha^{-1}$ $year^{-1}$, with a maximum of 15.78 Kg $Ha^{-1}$ $year^{-1}$. The highest mean value was registered in site IV (Instituto Tecnológico de León at SE) (**Figure 10**). From **Figure 10**, it can be observed that the mean ammonium deposition flux did not

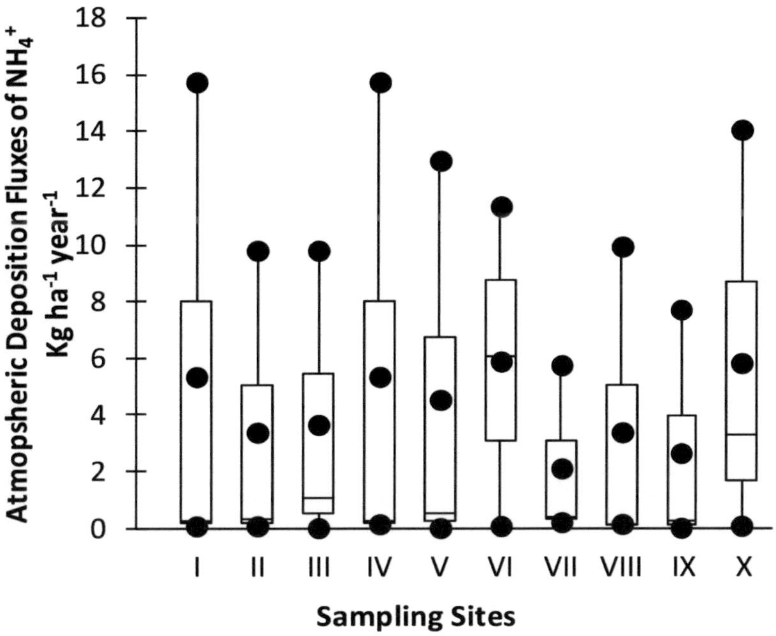


**Figure 9.**
*Ammonium atmospheric deposition fluxes by site.*

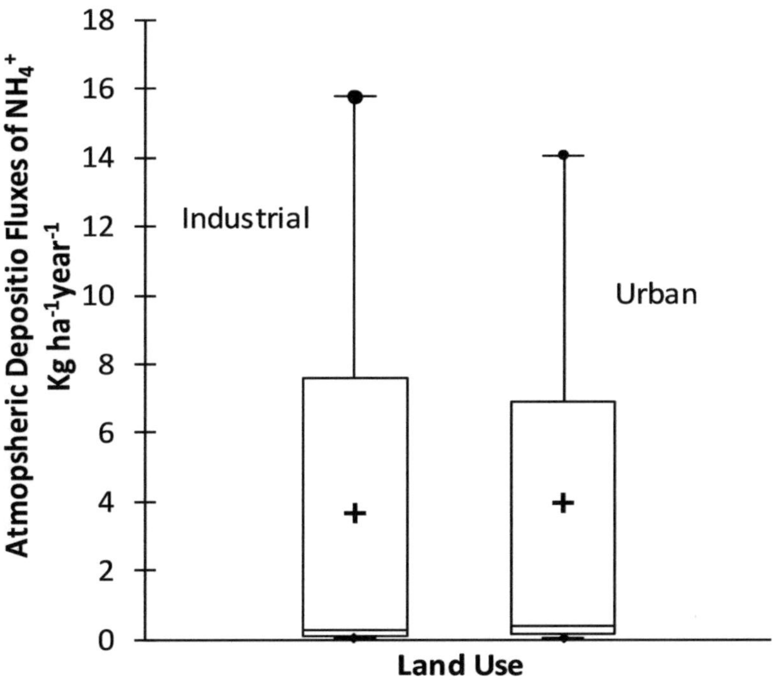


**Figure 10.**
*Ammonium atmospheric deposition fluxes by land use.*

show any variability or trend with respect to land use; however, the extreme values were higher in those sites with an industrial land use; however, when Friedman test was applied, any significant difference was found.

### 3.4 Meteorological analysis

To carry out the meteorological analysis at regional level, 24-h backward air-masses trajectories at 500, 1000, and 1500 m of altitude were obtained (a total of 85 trajectories) from HYSPLIT Model (National Oceanic and Atmospheric Administration), and choosing as representative site to CICEG-BOMBEROS, located at SE of the metropolitan area. In addition, meteorological surface data were obtained from SINAICA (National System of Air Quality Information) to estimate wind roses to identify the prevailing wind direction during the study period.

From surface analysis, it could be observed that wind direction showed a great variability between climatic seasons, for example, the prevailing wind direction during cold dry season was from SW (**Figure 11a**), during warm dry season was from NE (**Figure 11b**), and during rainy season was from E-NE (**Figure 11c**).

### 3.5 Mapping and reference values

*3.5.1 Mapping*

*3.5.1.1 $NH_4^+$*

In **Figure 12**, it can be observed that during warm dry and rainy seasons, the lowest ammonium deposition fluxes were registered, unlike cold dry season, when the highest levels were observed. During all study period, the highest ammonium

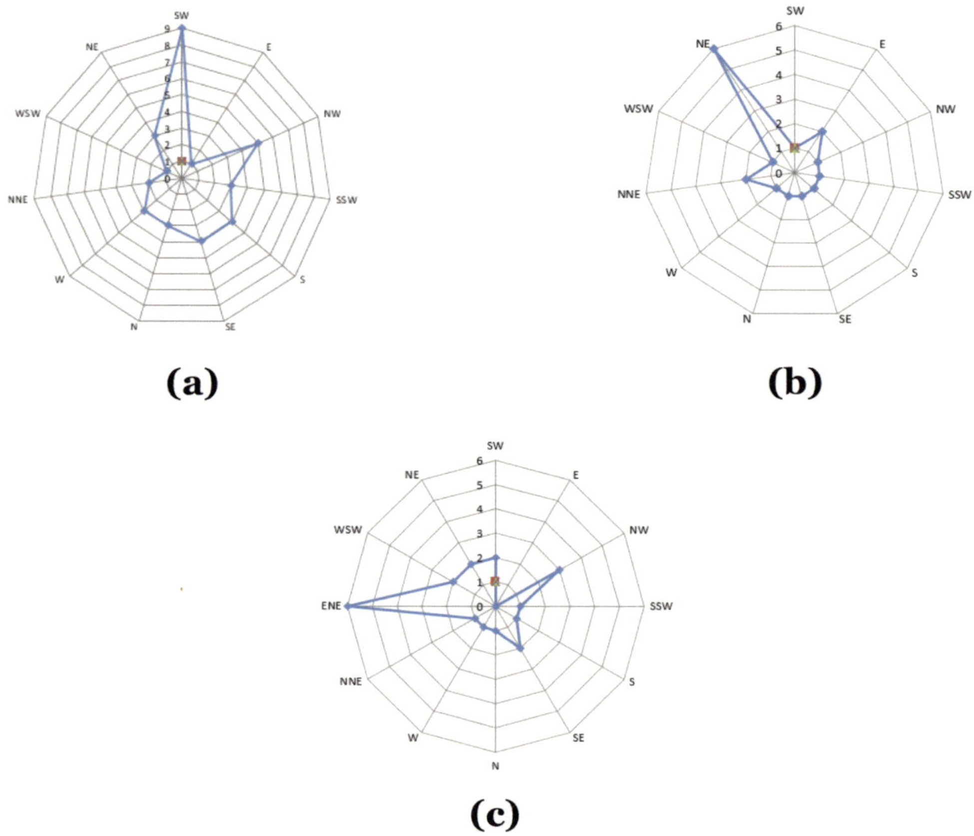


**Figure 11.**
*Frequency histogram for (a) cold dry season, (b) for warm dry season, and (c) for rainy season.*

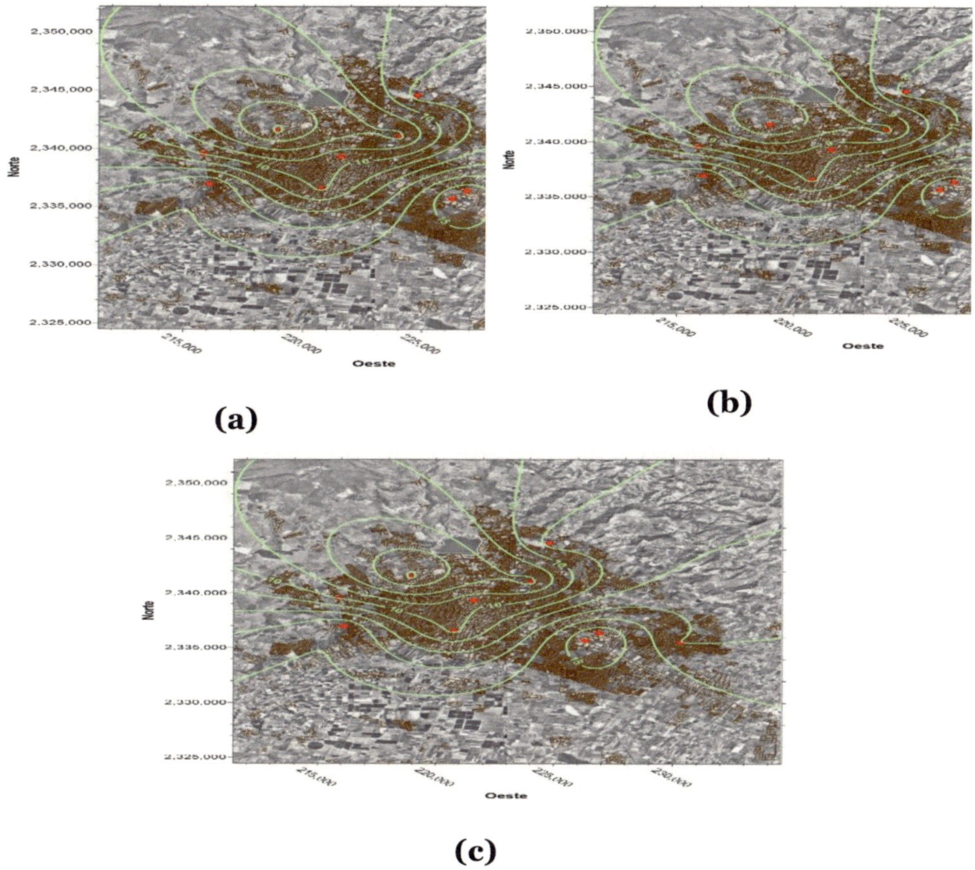


**Figure 12.**
*Deposition map of $NH_4^+$ (kg ha$^{-1}$ year$^{-1}$) in León City, (a) cold dry season, (b) warm dry season, and (c) rainy season.*

atmospheric deposition fluxes were found at the southeast of the city (CICEG-Bomberos and Instituto Tecnológico de León, both with an industrial land use). It is worth mentioning that both sites are influenced by the emissions released by a great vehicular fleet that circulates daily on the streets, especially near monitoring station CICEG-Bomberos which is located on the main route of entry to the city from Guanajuato and Silao.

#### *3.5.1.2 $NO_3^-$*

In the case of nitrate, from **Figure 13**, it can be observed that the highest nitrate deposition fluxes were found during the cold dry season. Regarding seasonal patterns, it can be observed that the highest nitrate deposition fluxes were registered at de Southwest of the city, specifically in site X, which corresponds to site of Loma Blanca. This zone has an urban land use with the highest altitude in the city (elevation of 1874 m), in which, there are many commercial areas and avenues with high vehicular traffic such as Miguel de Cervantes Av., Miguel Torres Landa Oriente Blvd. and Mariano Escobedo Oriente Blvd.

Since nitrate deposition fluxes were higher during the cold dry season, it suggests the origin of nitrate and its precursor in local sources. In addition, wind velocities are lower during the cold dry season in comparison with the other two seasons; therefore, there is a minor dispersion of pollutants in the region.

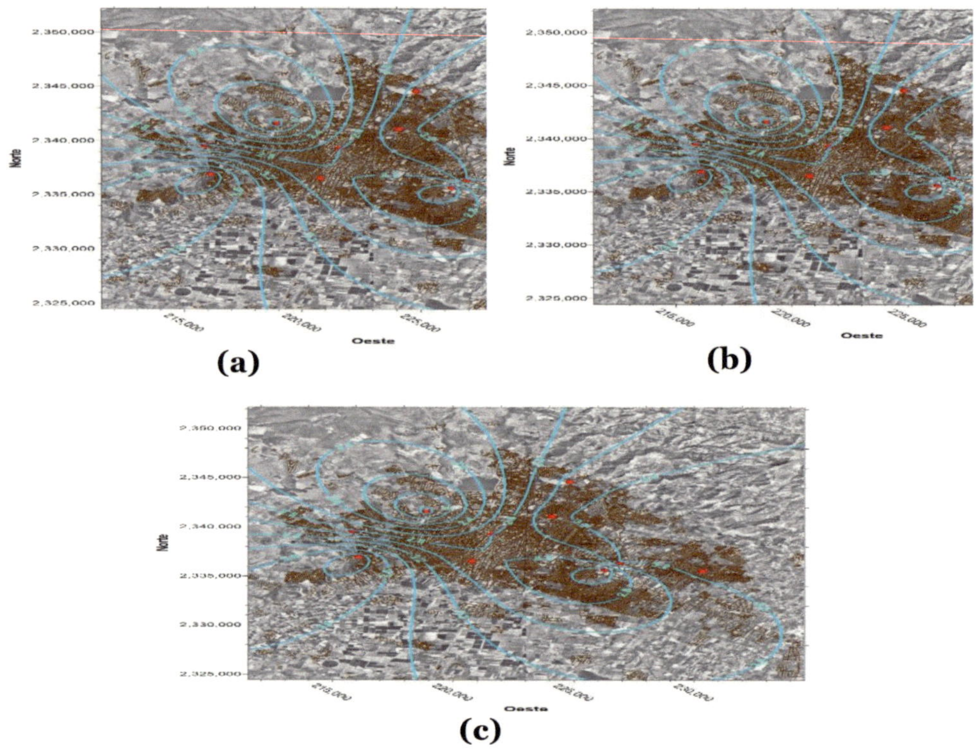


**Figure 13.**
*Deposition map of $NO_3^-$ (kg $ha^{-1}$ $year^{-1}$) in León City, (a) cold dry season, (b) warm dry season, and (c) rainy season.*

Another important factor is the lack of rain, contributing to a minor dilution or washing of the atmosphere. Levels found were homogeneously distributed as a result of local sources, and it agrees with residence time in the atmosphere of $NO_2$.

### *3.5.1.3 $SO_4^{2-}$*

In the case of $SO_4^{2-}$, its deposition fluxes were uniformly distributed in the metropolitan area of León (**Figure 14a–c**). In all seasons, the highest levels were found at NE (Site VI: Zoológico) and at SE (Sites I and IV corresponding to CICEG Bomberos and Instituto Tecnológico de León), all of them, with an industrial land use. On the other hand, the lowest sulfate deposition fluxes were found at North of the city.

The highest value for sulfate atmospheric deposition flux was found during the rainy season; however, this difference was not statistically significant in comparison with the other two seasons. Some hot spots were identified: at NE (Site VI), at SE (sites I and IV), and at SW (Site X).

### *3.5.2 Reference values*

It has been proposed a critical load value of 5 Kg N $ha^{-1}$ $year^{-1}$ for alpine ecosystems, which are more sensitive than ecosystems in lowlands [17], whereas in Nuevo México and California, a critical load value of 3–8 and 4–7 Kg N $ha^{-1}$ $year^{-1}$, respectively, has been proposed [18]. Regarding to Mexico, reference values to compare N and S atmospheric deposition fluxes are not available. On the other hand, for S, it has been proposed a critical load value of 3 Kg S $ha^{-1}$ $year^{-1}$ for sensitive areas in Europe, whereas for natural forests, values between 2 and 5 Kg S $ha^{-1}$ $year^{-1}$ have been proposed [18].

In this study, mean deposition fluxes for N (as $N\text{-}NH_4^+$ + $N\text{-}NO_3^-$) and S (as $SO_4^{2-}$) in metropolitan area of León were 5.82 N and 13.77 Kg S $ha^{-1}$ $year^{-1}$, respectively. In the case of N, exceeds the value reported for alpine ecosystems, but is near the reference values for Nuevo México and California. N deposition levels found in metropolitan area of León were twice as that reported by Cerón et al. [19] in Carmen Island, Campeche (2.15 N Kg $ha^{-1}$ $year^{-1}$); by Cerón et al. [20] in Orizaba Valley, Veracruz (1.44 N Kg $ha^{-1}$ $year^{-1}$); and by López [21] in Mérida, Yucatán (2.7 N Kg $ha^{-1}$ $year^{-1}$); also exceeding those values reported by Cerón et al. [22] in metropolitan area of Monterrey (4.88 N Kg $ha^{-1}$ $year^{-1}$) and five times as that reported by Cerón et al. [23] in Atasta-Xicalango, Campeche (1.15 Kg $ha^{-1}$ $year^{-1}$).

With respect to S, deposition fluxes exceeded almost 4.5 times the critical load value proposed for sensitive areas in Europe, whereas for natural forests, reference value was exceeded almost three times those values reported by Cerón et al. [19] in Carmen Island, Campeche (4.7 S Kg $ha^{-1}$ $year^{-1}$) and by López [21] in Mérida, Yucatán (4.07 S Kg $ha^{-1}$ $year^{-1}$), and twice the value reported by Cerón et al. [23] in Atasta-Xicalango, Campeche (8.57 S Kg $ha^{-1}$ $year^{-1}$). However, values reported by Cerón et al. [20] in Orizaba Valley (55.16 S Kg $ha^{-1}$ year^−1) and by Cerón et al. [22] in metropolitan area of Monterrey (25.03 Kg $ha^{-1}$ $year^{-1}$) were not exceeded. Therefore, it can be inferred that, N and S atmospheric deposition fluxes constitute a potential risk of acidification for ecosystems in the region.

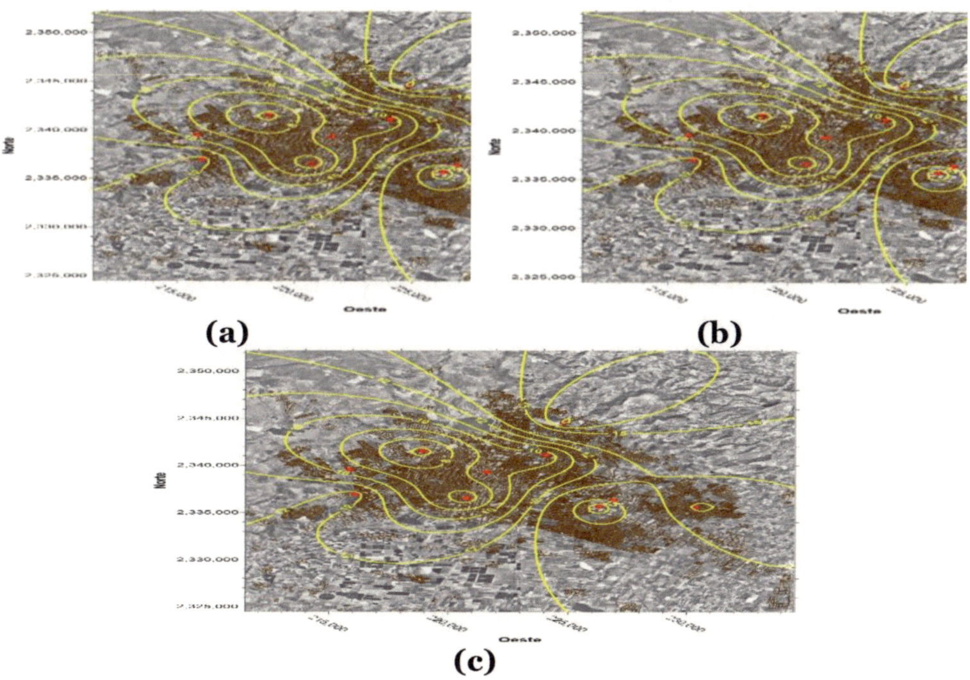


**Figure 14.**
*Deposition map of $SO_4^{2-}$ (kg $ha^{-1}$ $year^{-1}$) in León city, (a) cold dry season, (b) warm dry season, and (c) rainy season.*

## 4. Conclusions

In general, a strong seasonality in both, N and S atmospheric deposition fluxes was observed. Regarding to spatial variability, it was found that, in each climatic season, fluxes stayed homogeneous and uniformly distributed as a result of meteorological conditions prevailing in each season, having a regional origin during the rainy season and a local origin during the rest of the year. $NH_4^+$ and $NO_3^-$ showed a similar spatial and temporal pattern in their deposition fluxes. According to the analysis by season, the highest N deposition flux was found during the cold dry season, followed by warm dry season and rainy season, respectively. It could be explained due to the occurrence of a minor dispersion of pollutants resulting from lower wind velocities, and the frequent occurrence of thermal inversions during the cold dry season. On the other hand, dilution phenomena could play an important role during the rainy season, resulting in lower fluxes during this season. From spatial analysis of N deposition fluxes, it can be concluded that the site with the highest N deposition was Loma Blanca (Site X). In general, all sites where the highest N deposition fluxes were found are urban sites located in the suburbs of the city, with high vehicular traffic, where some agricultural sites are located. It suggests that the main sources of $NO_3^-$ are vehicular emissions, whereas the main source of $NH_4^+$ is the intensive use of fertilizers in these agricultural zones. However, since significant differences were not found between sampling sites by land use, it can be concluded that fluxes could be uniformly distributed, confirming their local origin. It agrees with the residence time of $NO_2$ in the atmosphere of 1 day, time in which, it is deposited as $NO_3^-$ or $NH_4^+$ at ground level, being their concentrations higher in the surroundings of the emission point. Since in León city, there was not a territorial ecological order in the past, the city growth and development along the years was not planned, resulting in the coexistence of different type of land use in a given zone.

On the other hand, S deposition fluxes showed a different pattern, suggesting their regional origin. From seasonal analysis, it was found that the highest atmospheric deposition flux occurred during the rainy season, followed by warm dry and cold dry seasons. It agrees with the regional character of $SO_2$ (sulfate precursor) which has a residence time in the atmosphere of 1–5 days [24, 25], being able to transport long distances before to be deposited finally at ground level. From air masses trajectories analysis, it could be concluded that, during the rainy season, this pollutant is long-range transported from the South of United States and from the North of Gulf of Mexico, explaining in this way, why during this season, a relative sulfate enrichment is observed in comparison with the other two seasons.

From analysis by sampling site, the site with the highest S atmospheric deposition flux was Instituto Tecnológico de León followed by Zoológico (sites IV and VI), these sites have an industrial and urban land use, suggesting that both vehicular and industrial emissions could contribute to background sulfate levels, which in turn were temporarily increased during the rainy season due to the long-range transport of air masses from distant sources (South of United States and North of Gulf of Mexico).

## Author details

Rosa María Cerón Bretón[1*], Julia Griselda Cerón Bretón[1], Reyna del Carmen Lara Severino[1], Marcela Rangel Marrón[1], María de la Luz Espinosa Fuentes[2], Simón Eduardo Carranco Lozada[3] and Lizbeth Cisneros Rosique[1]

1 Facultad de Química, Universidad Autónoma del Carmen, Ciudad del Carmen, Campeche, México

2 Centro de Ciencias de la Atmósfera, Universidad Nacional Autónoma de México, Ciudad de México, México

3 Instituto Politécnico Nacional, Guanajuato, México

*Address all correspondence to: rosabreton1970@gmail.com

## References

[1] Shahin U, Lu J, Yi SM, Paode RD, Holsen TM. Long-term elemental dry deposition fluxes measured around Lake Michigan with an automated dry deposition sampler. Environmental Science and Technology. 2000;**34**: 1887-1892. DOI: 10.1021/es9907562

[2] Kulshrestha UC, Kulshrestha MJ, Sekar R, Sastry GSR, Vairamani M. Chemical characteristics of rainwater at an urban site of south-Central India. Atmospheric Environment. 2003;**37**:3019-3069. DOI: 10.1016/S1352-2310(03)00266-8

[3] Laj P, Klousen J, Bilde M, Plab C, Pappalardo G, Clerbaux C, et al. Measuring atmospheric composition change. Atmospheric Environment. 2009;**43**:5351-5414. DOI: 10.1016/j.atmosenv.2009.08.020

[4] Fenn ME, Bytnerowicz A. Summer throughfall and winter deposition in the San Bernardino Mountains in southern California. Atmospheric Environment. 1997;**31**:673-683. DOI: 10.1016/S1352-2310(96)00238-5

[5] Warfvinge P. Integrating modeling. In: Heij GJ, Erisman JW, editors. Acid Atmospheric Deposition and Its Effects on Terrestrial Ecosystems in the Netherlands: The Third and Final Phase. Elsevier; 1997

[6] Lovland G, Andersen B, Joffre S, Pedersen U, Hormand M, Reissell A. Mapping deposition of sulphur, nitrogen, and base cations in the Nordic Countries. Swedish Environmental Research Institute. Report B 1055. 1992-06-25; 1992

[7] INEGI Prontuario de información geográfica municipal de los Estados Unidos Mexicanos León, Guanajuato. Clave Geoestadística. 2009; 11020:2

[8] INEGI, Población Total en Viviendas de la Ciudad de León, Guanajuato. 2015. Available from: https://www.inegi.org.mx/app/areasgeograficas/?ag=11

[9] Garten CT, Schwab AB, Shirshac TL. Foliar retention of $^{15}N$ tracers: Implications for net canopy exchange in low- and high-elevation forest ecosystems. Forest Ecology and Management. 1998;**103**:211-216

[10] Fenn ME, de Bauer LI, Quevedo-Nolasco A, Rodríguez-Frausto C. Nitrogen and sulfur deposition and forest nutrient status in the valley of Mexico. Water, Air, and Soil Pollution. 1999;**113**:155-174

[11] Fenn ME, Poth MA. Monitoring nitrogen deposition in throughfall using ion exchange resin columns: A field test in the San Bernardino Mountains. Journal of Environmental Quality. 2004;**30**:2007-2014. DOI: 10.2134/jeq2004.2007

[12] Clow D, Roop H, Nanus L, Fenn M, Sexstone G. Spatial patterns of atmospheric deposition of nitrogen and sulfur using ion-exchange resin collectors in Rocky Mountain National Park, USA. Atmospheric Environment. 2015;**101**:149-157. DOI: 10.1016/j.atmosenv.2014.11.027

[13] Fresenius W, Quentin KE, Schneider W. Water Analysis: A Practical Guide to Physico-Chemical, Chemical and Microbiological Water Examination and Quality Assurance. New York, NY: Springer-Verlag; 1992. pp. 420-421

[14] Secretaría de Economía. Norma Mexicana. NMX-AA-074-SCFI-2014. Medición del ión sulfato en aguas naturales, residuales y residuales tratadas. Método de prueba. Dirección General de Normas. 2015

[15] Secretaría de Economía. Norma Mexicana. NMX-AA-079-SCFI-2001.

Análisis de Aguas. Determinación de nitratos en aguas naturales, potables, residuales y residuales tratadas. Método de prueba. Dirección General de Normas. 2000

[16] Sáez J. Investigación Educativa, Fundamentos teóricos, Procesos y Elementos prácticos, Enfoque práctico Con Ejemplos [thesis]. Madrid: Universidad Nacional de Educación a distancia; 2017

[17] Fenn ME, Geiser LH. Temperate Sierra. In: Pardo LH, Robin-Abbot MJ, Driscoll CT, editors. Assessment of Nitrogen Deposition Effects and Empirical Critical Loads of Nitrogen for Ecoregions of the United States. General Technical Report NRS-80. Delaware, OH: United States Department of Agriculture; 2011. pp. 175-180

[18] Grennfelt O, Nilsson J. Critical Loads for Sulfur and Nitrogen. Report from a Workshop held at Skokloster, Sweden; 19-24 March 1988. Copenhagen, Denmark: The Nordic Council of Ministers Report; 1988. p. 15. ISBN: 87-7303-248-4

[19] Cerón R, Cerón JG, Muriel M, Anguebes F, Ramírez M, Zavala JC, et al. Spatial and temporal distribution of throughfall deposition of nitrogen and sulfur in the mangrove forests associated to Terminos lagoon. In: Farhad N, editor. Current Air Quality Issues. IntechOpen, Rijeka, Croatia; 2015. pp. 147-164. DOI: 10.5772/59726

[20] Cerón R, Cerón J, Kahl J, Lara R, Rangel M, Rustrián E. Influencia potencial de las emisiones y el transporte de N y S en el Valle de Orizaba, Veracruz, México. Agrociencia. 2018;**52**:653-670

[21] López D, Cerón J, Cerón R, Rangel M, Uc M, Salvaño Y, Lara R. Mapeo del depósito atmosférico de azufre y nitrógeno durante la primavera de 2017 en la Zona Metropolitana de la Ciudad de Mérida. In: Proceedings of VII Congreso Nacional de Investigación en Cambio Climático. Centro de Ciencias de la Atmósfera. UNAM; 3-4 October 2017; Mexico City

[22] Cerón RM, Cerón JG, Kahl J, Ramírez E, Córdova A, Espinosa A, et al. Atmospheric N and S deposition fluxes in the metropolitan area of Monterrey, Mexico and its relation with criteria air pollutants and meteorological conditions. In: Jorge Del Real Olvera, editor. Air Pollution-Monitoring, Quantification, and Removal of Gases and Particles. London, UK: IntechOpen; 2018. pp. 75-98. DOI: 10.5772/intechopen.79484

[23] Cerón J, Cerón R, Aguilar C, Montalvo C, García A, Muriel M, et al. Mapping temporal and spatial variation of sulphur and nitrogen deposition to a complex ecosystem in Campeche, Mexico. Transactions on the Ecology and the Environment. 2016;**203**:113-123. Environmental Impact II. UK: WIT Press. DOI: 10.2495/EID160111

[24] Kahl J, Bravo A, Soto A, Sosa E, Sánchez A, Alarcón A. Characterization of atmospheric transport to the El Tajín archeological zone in Veracruz, México. Atmosfera. 2007;**20**(4):359-371

[25] Cerón R, Kahl J, Cerón J, Solís JA. The influence of meteorology and atmospheric transport patterns on the sulfate levels in rainwater in Orizaba Valley, Veracruz, Mexico. Journal of Environmental Protection. 2019;**10**(6):1-17. DOI: 10.4236/JEP.2019.106049

Section 2

# Geological Sciences

Chapter 4

# Monitoring Storm Impacts on Sandy Coastlines with UAVs

*Alex Smith, Brianna Lunardi, Elizabeth George and Chris Houser*

## Abstract

UAV applications have shown the potential to increase the efficiency of collecting high resolution and spatially extensive topographic datasets of sandy coastal systems. These systems are dynamic and sensitive to variability in wave energy, evident in topographic adjustments associated with storm events. Topographic and volumetric changes of a beach-dune system were measured following a post tropical storm event. Using a pre-storm LiDAR and post-storm UAV survey, we identified high magnitude and continuous alongshore erosion of the foredune. Lower magnitude and discontinuous areas of deposition were also recorded, as sediment eroded from the foredune translated seaward and was deposited onto the beach. Overall, a total volumetric loss of ~11,000 $m^3$ from the beach-dune zone was recorded along the 2.5 km survey extent. Our results highlight the capability of UAVs for rapid monitoring and quantification of storm impacts. Furthermore, confidence in reported topographic changes was improved by implementing quality control measures and handling of data uncertainties (e.g., vegetation). The aim of this chapter is to quantify the impact of a storm event on a beach-dune system and discuss methodological challenges of monitoring sandy coastlines with UAVs.

**Keywords:** UAV applications for coastal monitoring, structure from motion, storm impacts, beach-dune interactions, dune recovery

## 1. Introduction

Sandy coastlines are dynamic environments that are continuously modified in response to wave, tidal, and eolian processes. Sediment is in constant flux amongst the nearshore, foreshore, and backshore (**Figure 1**). Over short time scales (i.e., hours to months), individual storm events and seasonal variability in wave and wind energy result in topographic adjustments [1]. For example, characteristic 'winter' or 'summer' profiles can develop where sediment movement between zones can be cyclically removed, stored, and/or returned [2, 3] (**Figure 1**). The 'winter' profile develops in periods of higher wave energy where sediment is removed from the backshore (i.e., through beach and foredune scarping) and stored in nearshore sand bars [2, 3]. As wave energy decreases, sediment can be returned landward through welding of nearshore sand bars onto the foreshore and deposition

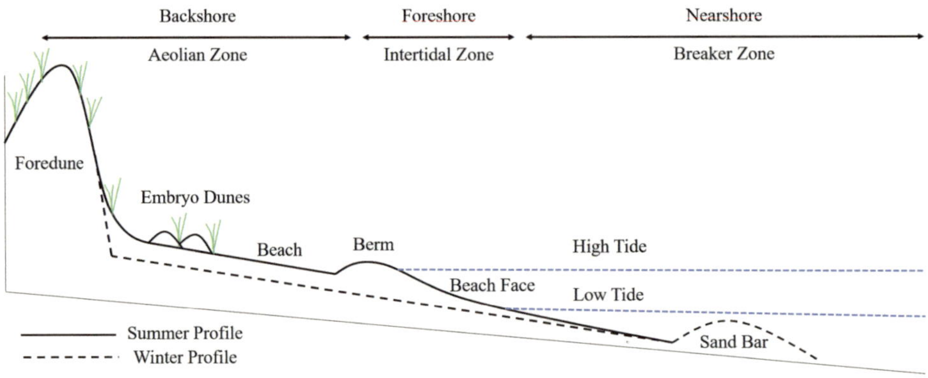


**Figure 1.**
*This idealized diagram depicts the seasonal variability of across shore profiles, landforms, and vegetation that can develop during periods of increased (i.e., winter profile) or decreased (i.e., summer profile) wave energy.*

of berm deposits on the beach [2, 3], leading to embryo dune development and foredune recovery [4] (e.g., 'summer' profile; **Figure 1**).

Small scale topographic variability is subsumed by larger scale controls on coastal dynamics [1, 5, 6] except in the presence of an alongshore variable framework geology [7]. However, changes in wave climate [8] and eustatic sea level [9] have the potential to disrupt the current balance between erosional and depositional processes. These climatic changes could lead to increased water levels during storm events and the potential for sediment to be transported inland through breaching [10, 11], overwash [12] or the development of blowouts [13, 14]. Sediment deposited landward of the foredune would be effectively removed from the cyclical 'seasonal' recovery state, and could lead to increased erosion, fragmentation, and landward retreat of the foredune [15, 16]. This disruption of sediment supply could also accelerate the transgression of the coastline as it responds to future changes in sea level [9]. Thus, the ability of foredunes to recover following storm events will have implications on both the short- and longer-term resiliency of sandy coastlines [17, 18].

Following significant storm events or storm seasons, periods of increased wave energy often leads to the development of a foredune scarp (**Figure 2**). The ability of the foredune to recover, or return to its pre-storm morphology, is then controlled primarily by the sequence and relationship between eolian transport potential and sediment supply [4]. Initially, sediment eroded from the foredune by elevated storm surge and wave run up can be deposited directly onto the beach or further seaward into nearshore bar structures [2, 3]. As the beach slope relaxes to a characteristic 'winter' or flattened beach profile (**Figure 2**), the low sloping surface promotes rapid boundary layer development during on shore winds and increases sediment flux potential [19], with minimal slope controlled limitations on flux magnitude [20]. Elevated wrack lines (**Figure 2**), exposed lag deposits, and large woody debris that may be present following storm events can temporarily trap sediment blown into the beach-dune boundary (e.g., see [6]).

Over seasonal or annual time scales, sediment stored in nearshore bars can begin migrating landward during periods of reduced wave energy [2, 3]. An increase in sediment supply into the foreshore can result in the formation of multiple berm ridges that are deposited above high tide and swash lines, increasing beach width [3]. This results in a larger supply of dry erodible sediment in the backshore which is less affected by tidal or swash driven surface moisture constraints on eolian transport [21]. Finally, vegetation recolonization at the beach-dune boundary can

**Figure 2.**
*A foredune scarp observed at Brackley Beach, Prince Edward Island National Park, following the post-tropical storm Dorian in September 2019. This image shows that significant erosion of the stoss slope resulted in the formation of a steep and continuous scarp, alongshore. Other storm impacts, including a flattened across shore profile and elevated wrack line, are also visible.*

limit transport potential and initiate dune building (e.g., [22]). If these conditions persist, dune recovery can occur through ramp building and embryo dunes can develop seaward of the established foredune [15]. Following major erosion events, full dune recovery may take years to complete [17, 23] and is controlled by frequency of high magnitude events [17].

The impact of a storm and the relatively slow recovery tend to be considered a two-dimensional, cross-shore phenomenon in which the storm impact depends only on the height of the storm surge relative to the elevation of the dune crest (e.g., [24, 25]). However, the foredune line is not uniform and can exhibit considerable variability in height, volume, and alongshore extent [12, 26, 27]. As noted, the exact location of dune erosion and overwash penetration depends on the correspondence of alongshore variations in the incident forcing and on existing gaps and low-lying areas along the dune line [28–31]. Understanding the variability of the beach-dune systems is essential to understanding the response of sandy coastlines to changes in storm activity and sea level rise, and it is important to the development of appropriate sampling strategies for field studies of sediment transport exchange amongst the nearshore, beach and dune (e.g., [31]).

Climatic change over the coming decades, including increased storminess [8] or sea level rise [9], have the potential to modify current beach-dune interactions. For example, an increased frequency of storm surge events can lead to barrier island systems of low elevation and discontinuous dunes that, in turn, increase the potential for island inundation and breaching [17]. However, a low frequency of storm surge can limit sediment transfer to the backbarrier as overwash leading to island drowning in response to an increase in sea level [32], unless sediment transfer is accomplished by blowouts [13, 14]. Monitoring the resiliency of sandy coastlines is, therefore, critical to our understanding on how these systems will respond to larger scale sedimentological and climatic perturbations. To address this challenge, advances in surveying technology including unmanned aerial vehicles (UAVs) and light detection and ranging (LiDAR) systems are able to provide robust geo-spatial data sets that will enhance repeat survey strategies of these dynamic environments.

### 1.1 Coastal applications for UAVs

The affordability and capability of UAV surveys to rapidly produce spatially extensive and high-resolution terrain models has been a boon to geoscientific research over the past decade (e.g., [33]). The majority of which has used Structure from Motion (SfM) photogrammetry. This technique generates automatic tie points (ATPs) between overlapping images and triangulates their position to produce a 3-D point cloud [34, 35]. The utility of UAVs and SfM have been well documented; however, difficulties remain in accurately capturing areas of topographic complexity, water boundaries, and vegetation [33]. The latter two can be particularly problematic in coastal environments due to variability in tidal ranges and/or wave run up on the seaward boundary [36, 37] and in the density of seasonally intermittent or established vegetation across the beach and foredune [38–41]. Other common problems that arise in coastal environments include wet or low texture surfaces that limit image recognition [41], maximum operational wind conditions of $\sim < 25$ km/h [37, 42], and regulations that restrict UAV flights in the vicinity of pedestrians [37].

Due to these difficulties, the breadth of UAV research monitoring sandy coastlines has been relatively limited. Research testing the quality of SfM derived point clouds to those produced by more traditional methods such as terrestrial laser scanning (TLS) has found good agreement in areas with limited vegetation height or density [39, 41]. Furthermore, the geo-referencing of UAV point clouds using survey ground control points (GCPs) have been able to obtain centimeter scale error over non-vegetated surfaces [36, 38–40, 43]. However, the elevational error over vegetated surfaces can be an order of magnitude or higher, ranging from the 10–100-cm scale and is largely controlled by vegetation density and canopy height [41]. To date, only a small number of UAV coastal monitoring studies have quantified topographic or volumetric changes of beach-dune morphology in response to storm events [37], storm seasons and human management [38, 44], or annual cycles [40, 45].

UAV and SfM applications have displayed the potential to increase the efficiency and rapidity of coastal monitoring. In order for this potential to be fully realized, difficulties arising during data collection, processing, and analysis must be addressed when developing systematic and repeatable survey strategies. Recently, UAV coastal research have also used multi-spectral (e.g., [46]) and LiDAR sensors (e.g., [47]); however, this chapter will focus specifically on UAV and SfM systems. The remainder of this chapter will be an introduction to basic quality control measures, handling of survey and environmental uncertainties, and reporting of topographic and volumetric changes associated with a storm event. Furthermore, the persistent challenges of coastal monitoring with UAVs and prospective solutions will be discussed. Ultimately, the aim of this chapter is to present a repeatable methodology to confidently report topographic change and highlight future methodological advances that are still needed to improve upon current coastal monitoring strategies using UAVs.

## 2. Study site

Brackley Beach (BB; 46°25′50″ N 63°11′50″ W) is a part of the Prince Edward Island National Park, located on the north shore of Prince Edward Island (PEI), Canada (**Figure 3A**). BB extends ~6 km alongshore in an east to west orientation and is backed by a highly continuous foredune. On September 7th and 8th BB was impacted by the post-tropical storm Dorian (SD). Sustained N–NNE winds of up to

**Figure 3.**
*Brackley Beach (A), Prince Edward Island National Park, extends ~6 km in an east to west orientation and is backed by highly continuous foredune. Following the storm Dorian, a flattened beach profile (B), significant foredune erosion (C), a continuous scarp developed alongshore (D and E).*

55 km/h and gusts of up to 93 km/h were recorded at the Stanhope meteorological station (SMS), located ~9 km east of BB. It must be noted that SMS records wind conditions at 3 m above the surface, or 7 m below standard meteorological recordings, and likely underrepresent the peak wind speeds associated with SD. Additionally, there are no offshore buoys on the north coast of PEI to record marine conditions associated with SD; however, significant wave heights of 7–8 m and a storm surge of 1.2 m was forecasted by the Dalcoast-HFX model [48]. Peak storm surge levels, coincident with a ~0.8 m high tide recorded on the north coast of PEI at 2 a.m. on September 8th, are estimated to be ~2 m above mean sea level (MSL). An initial site assessment of SD's impact on BB observed modification of the beach-dune morphometry including a flattened beach profile (**Figure 3B**), scarping of the entire frontal slope (i.e., from the dune toe to dune crest) of the foredune at the eastern section of BB (**Figure 3C**), and a continuous 1–2 m scarp across the majority of the mid and western sections of BB (**Figure 3D** and **E**).

## 2.1 Pre-SD baseline survey

An integrated aerial LiDAR topographic and bathymetric (topobathy) survey was conducted at Prince Edward Island National Park between the 4th and 7th of July 2019. The LiDAR data was obtained through CBCL Limited with permissions from Parks Canada, as part of the Federal Transportation Risk Assessment Initiative. A Leica Chiroptera II dual sensor Topobathy LiDAR, utilizing an infrared (240 kHz) and green laser (35 kHz), seamlessly captured 291 m swaths of both the topographic surface and near shore bathymetry, up to 5 m water depth. The survey was flown at an altitude of 400 m above the ground surface and maintained a sampling density of 2.72 points per meter (p/m) for the bathymetric and 18.6 p/m for the topographic surfaces. Point clouds were then geo-referenced using a total of 141 ground control points with a root mean square error (RMSE) of 0.05 m for the vertical transformation. The point cloud data was then classified for surficial and supra-surficial elements (e.g., vegetation, water surface, etc.). Following classification of the point cloud, only bathymetric and topographic surface classes were maintained to produce a 1 m $\times$ 1 m 'bare earth' or digital terrain model (DTM) that will serve as a pre-SD baseline for comparisons to post-SD UAV surveys.

## 3. Methodology

### 3.1 UAV survey design and initial SfM quality controls

On the 19th and 20th of September 2019, 11 days following SD, a survey consisting of eight overlapping UAV flight grids (**Figure 4A**) was flown at BB. Images were collected with a Mavic 2 Pro quadcopter and covered an alongshore extent of ~2.5 km. Across shore, UAV flight paths were programmed to capture images from the nearshore, foreshore, backshore, and back dune zones with 80% frontal- and 70% side-overlap. The UAV was flown at an altitude of 50 m corresponding to a ground sampling distance of ~1 cm/pixel. In each flight grid, at least 10 bright white bucket lids (30 cm in diameter) were used as GCPs and spread in a non-uniform placement across the foreshore, beach, and foredune (**Figure 4A**). The geographic location of each GCP was surveyed using a Global Navigation Satellite Series GPS. The high number of GCPs was designed to increase the accuracy of the geographic coordinate conversion by providing a diverse set of elevational references and to systematically build in redundancy that allow for GCP

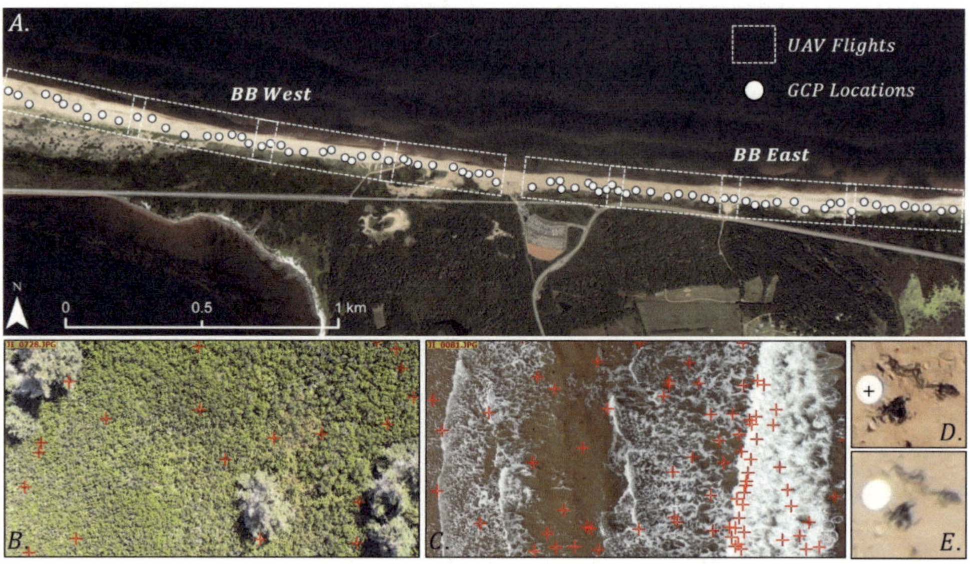


**Figure 4.**
*UAV flight and GCP locations for the Brackley Beach (BB) east and BB west surveys (A). Quality control measures were taken during post-processing including removing photos that generated a limited number of ATPs, typically occurring in areas of homogenous vegetation or prominent peaked vegetation (B) and breaking waves in the nearshore (C). Also, only GCPs that were clearly visible and non-deformed were selected (D) while all others were left unmarked (E) in order to improve accuracy during geo-referencing.*

removal if elevated error values are recorded. Common causes of error can include GPS survey (e.g., limited number of satellites or line of sight), pedestrian interference during UAV flights, or user identification error during post processing.

Images of all connected flight grids were collectively processed (i.e., BB West and BB East) using the commercial SfM software Pix4D. Alternative SFM software including Agisoft Metashape [36, 39, 40, 42–45] and Fledermaus v-7 [38] have been previously used and described. The remainder of this section will focus on the general Pix4D workflow used in this study. Initially, a target number of 10,000 ATPs were generated from each overlapping image. Next, Automatic Aerial Triangulation (AAT), Bundle Block Adjustment (BBA), and camera calibration were optimized for all images. Uncalibrated cameras, a result of errors in the internal (e.g. vibrations) or external (e.g., position and orientation) camera parameters, were deactivated to remove potential topographic deformation during point cloud generation.

Furthermore, flight lines on the periphery of the survey grid had difficulty finding a sufficient number of ATPs due to areas of homogenous vegetation (**Figure 4B**), selecting prominent features such as treetops from multiple angles (**Figure 4B**), or breaking waves in the nearshore (**Figure 4C**). As a result, severe over- or under-estimation of the surface can occur by misclassifying reference elevation values. In an attempt to reduce survey error, images were clipped to the primary area of interest (i.e., the beach-dune zone). This ensured an optimal number of ATPs present in each image and removed potential edge or 'bowl' effects that can distort the point cloud in areas away from the GCP locations [40]. Next, GCPs that were clearly visible in the flight images were zoomed into and marked at their centroid (**Figure 4D**) to ensure proper pixel selection. GCPs that were blurry (**Figure 4E**), warped, displayed limited contrast, or not entirely visible were not selected because un-proper pixel selection can also introduce error during geographic transformation. Marked GCPs were then used for geographic conversion resulting in a vertical RMSE of 0.028 and 0.021 for BB West and East, respectively.

### 3.2 DSM and DTM generation and environmental uncertainty

Following initial processing, a point cloud was generated and consisted of $2.65 \times 10^{-8}$ points with a density of $\sim$2000 points per $m^2$. The point cloud was classified using Pix4D's predefined class groups including ground, road surface, high vegetation, building, and human made objects, in order to improve DTM filtering. Point interpolation was completed using an inverse distance weighting (IDW) approach to generate a universal DSM (i.e., retaining all elevational classes) and filtered DTM (i.e., retaining only the ground elevation class) at a 1 m $\times$ 1 m resolution for direct comparisons to the pre-SD LiDAR DTM.

Prior to measuring topographic changes between surfaces, an initial assessment of the quality of the LiDAR DTM and UAV DSM and DTM displayed significant irregularities (**Figure 5A**). For instance, the LiDAR DTM was generated using green laser that can penetrate up to 5 m depth and full wave form infrared laser that can penetrate the vegetation canopy. This provides a fully integrated bathymetric and topographic surface transitioning from the nearshore—to backshore zones (**Figure 5A**). SfM is not able to penetrate the water column and has difficulty measuring ground points in vegetated areas. As a result, the SfM DSM captures noise above the surface that is associated with wave breaking and run up in the nearshore and foreshore zones (**Figure 5A** and **B**). In the backshore, the vegetated crest and lee slope is overestimated on average by $\sim$0.5–1 m (**Figure 5A**) and represents variability in vegetation density and canopy height (**Figure 5B**).

Alternatively, the standard Pix4D DTM filtering method almost completely removed the foredune in areas that recorded significant scarping. **Figure 5A** shows

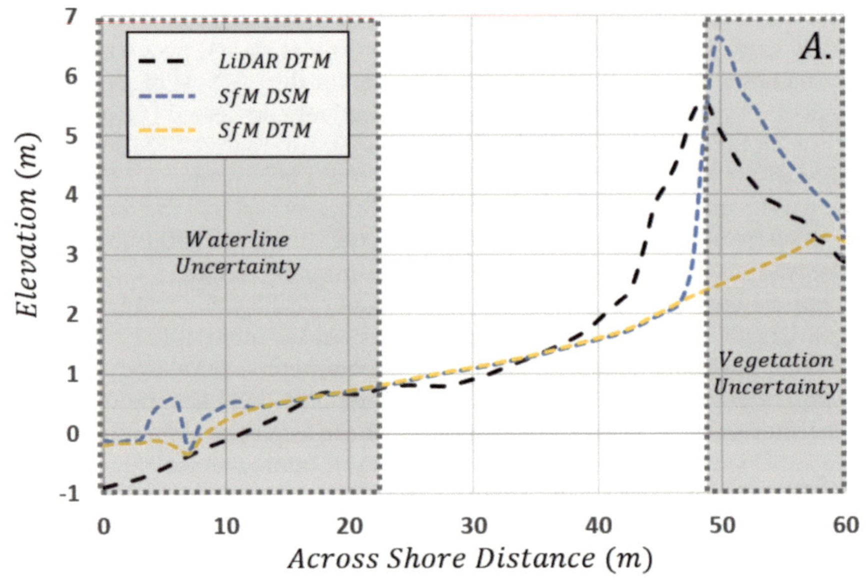

**Figure 5.**
*Across shore profiles comparing the pre-storm Dorian (SD) LiDAR DTM and post-SD UAV DSM and DTM (A). Inconsistencies occur in the UAV surveys due to over-estimation of elevation values due to environmental uncertainties including the waterline (A) in the intertidal zone (B) and vegetation (A) in the vegetated back dune zone (B). Also, under-estimation of elevation values occurred due to over-filtering of the UAV DTM has almost entirely removed the foredune following SD.*

a profile that was taken from a highly eroded foredune at BB East. Notice, only a small crest remains visible more than 2 m below and 10 m landward of the dune crest measured by the LiDAR DTM (**Figure 5A**). The removal of the foredune from the UAV DTM is indicative of over-filtering, not storm erosion. While good agreement exists between the SfM DTM and DSM over the non-vegetated upper beach surface (**Figure 5A** and **B**), the DTM filtering method could not resolve significant breaks in slope. Therefore, our UAV DTM was not capable of quantifying topographic changes associated with a distinct scarped foredune and will not be used for our post storm measurements.

To limit the environmental uncertainty associated with the UAV DSM, and to create a repeatable survey methodology, topographic change measurements must be confined to areas not affected by potential error introduced by water or vegetation. Difficulty arises when demarcating a seaward boundary due to the fluctuation of the water line in the intertidal zone (**Figure 5B**). The higher high water (HHW) line, 0.8 m above mean sea level MSL [49], or average of the highest annual tide levels was chosen as the seaward boundary. The HHW line is typically above the fluctuating water line and provides a repeatable method to measure spatially comparable topographic changes through time, regardless of the yearly, monthly, or daily variability in tide ranges and cycles. The landward extent of the DSM was limited to the foredune scarp because it marks the boundary between non-vegetated and vegetated surfaces, post-SD (**Figure 5B**). By removing significant environmental uncertainties associated with the water and vegetation line, we improve our confidence in monitoring topographic changes occurring between the pre-SD LiDAR DTM and post-SD UAV DSM.

### 3.3 Volumetric change and uncertainty

Volumetric changes ($\Delta V$ (m$^3$); Eq. (1)) from pre- and post-SD are reported within the detectable range of the SfM derived DSM. Here, $z_2$ and $z_1$ represent the elevation of the surface in time series two (i.e., post-SD) and time series one (i.e., pre-SD), respectively, and $x$ $(m)$ and $y$ $(m)$ are the length and width of the raster pixels $(p)$. To assess the accuracy of the $\Delta V$ measurements, the propagated error ($PE$ (m)); Eq. (2)) reports the magnitude of error ($e(m)$) associated with both time series. Common $e$ metrics include the RMSE [36, 38–40, 43, 44] and standard deviation error ($\sigma$, [50, 51]). Once the $PE$ is determined, the volumetric change uncertainty ($\Delta VU$ (m$^3$); Eq. (3)) provides a universal error value associated with each $p$ and allows for a range of uncertainty (i.e., $\pm\Delta VU$) to be reported with the $\Delta V$ measurements (e.g., [50, 51]).

$$\Delta V = (z_2 - z_1)xy \tag{1}$$

$$PE = \sqrt{e_1^2 + e_2^2} \tag{2}$$

$$\Delta VU = n(p)PE \tag{3}$$

To minimize the uncertainty of the $\Delta V$ measurements, a threshold is applied to remove values of low magnitude topographic change. The magnitude of change, away from (i.e., $\pm$)$PE$, is measured using a statistical t-score approach ($t$; Eq. (4); [52]). A minimum 95% confidence level ($CL_{95\%}$; Eq. (5)) threshold of ~1.96, valid for large population sizes, is determined based on a two-tailed test that accounts for both negative and positive values that correspond to erosion and deposition, respectively. Absolute $t$ values that exceed 1.96 are preserved representing areas of low uncertainty, while $t$ values less than 1.96 are classified as zones of high uncertainty and removed from $\Delta V$ measurements.

$$t = \frac{z_2 - z_1}{PE} \tag{4}$$

$$CL_{95\%} = |t| \begin{cases} for & |t| > 1.96 \\ else & \varnothing \end{cases} \tag{5}$$

## 4. Results

Topographic changes, measured between the pre-SD LiDAR DTM and post-SD UAV DSM, display high magnitude and continuous erosion that mark the foredune scarp along BB (**Figure 6A** and **B**). The scarp line consistently eroded into the seaward base of the foredune by >1 m, with higher magnitude erosion of the frontal section of the foredune of over 4 m recorded at eastward extent of the BB East survey (**Figure 6B**). Relatively low magnitude deposition of sediment of 0.2–0.4 m is observed in discontinuous areas of the beach, becoming more prominent at BB west. Despite these areas of significant erosion or depositional changes, 42% of BB (**Figure 6A** and **B**) is classified as a zone of uncertainty (i.e., below the $CL_{95\%}$ threshold). Low magnitude topographic changes of $>-0.11$ and $<0.11$, corresponding to a $|t|<1.96$, often mark an area of transition between sediment eroded from the foredune and deposited onto the beach. This area of transition or slope relaxation is typical of a 'winter' or flattened beach profile following storm events and was observed over the entirety of BB (e.g., **Figures 2** and **3B**).

Post-SD, a total volume change ($\Delta V$) of $-11{,}004$ m$^3$ and volume change uncertainty ($\Delta VU$) of $\pm 4704$ m$^3$ was recorded (**Table 1**). High magnitude erosion, accounting for 49% of the detectable surface (**Table 1**), is the primary geomorphic change driver; however, this is associated with significant $\Delta VU$ accounting for 42% of the cumulative volume change value. To reduce the uncertainty associated with low magnitude elevation changes, the topographic change threshold ($CL_{95\%}$) was applied. As a result, $\Delta V$ remained similar at $-11{,}323$ m$^3$ but $\Delta VU$ was reduced to $\pm 2659$ m$^3$ (**Table 1**). Notice that the higher magnitude erosional values now represents 57% of the $CL_{95\%}$ surface and $\Delta VU$ is reduced, accounting for only 23% of the cumulative volume change. Furthermore, after applying the topographic change threshold the observational area was reduced by 43% and $\Delta VU$ by 54% (**Table 1**). By accounting for the uncertainty introduced by the survey error, and associated with low magnitude elevation changes, we have systematically reduced $\Delta VU$ and improve our confidence in reported volume changes post-SD.

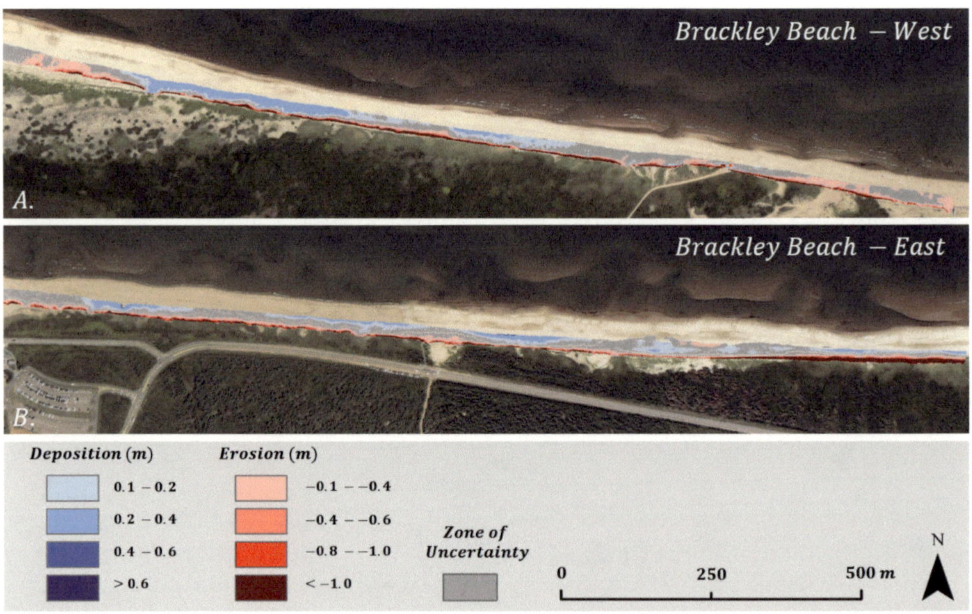


**Figure 6.**
*Elevation changes at Brackley Beach West (A) and East (B), measured from the pre-SD LiDAR DTM and post-SD UAV DSM. A high magnitude and continuous erosional scarp is visible (i.e., in dark red) along the beach-dune boundary with the highest magnitude change occurring at the eastern extent of Brackley Beach East (B). Lower magnitude deposition (i.e., in blue) was observed to develop intermittently on the upper beach surface, becoming more continuous at Brackley Beach West (A).*

| | Observed | $CL_{95\%}$ |
|---|---|---|
| Erosion ($m^3$) | −15,975 ± 2349 | −15,201 ± 1509 |
| Area ($km^2$) | 0.038 (49%) | 0.025 (57%) |
| Deposition ($m^3$) | 4971 ± 2354 | 3878 ± 1149 |
| Area ($km^2$) | 0.039 (51%) | 0.019 (43%) |
| Total change ($m^3$) | −11,004 ± 4704 | −11,323 ± 2659 |
| Total area ($km^2$) | 0.077 | 0.044 |

**Table 1.**
*Observed and threshold ($CL_{95\%}$) volume change ($\Delta V$), volume change uncertainty ($\Delta VU$), and area are reported for all erosional, depositional, and total change surfaces.*

### 4.1 Alongshore variability

Alongshore, the eastern extent of the BB survey recorded the highest magnitude topographic change. This is particularly evident in areas that experienced erosion of the full-frontal section of the foredune (**Figure 7A** and **B**). In these locations, the scarp line forms at or behind the former crest line, leading to slope failure,

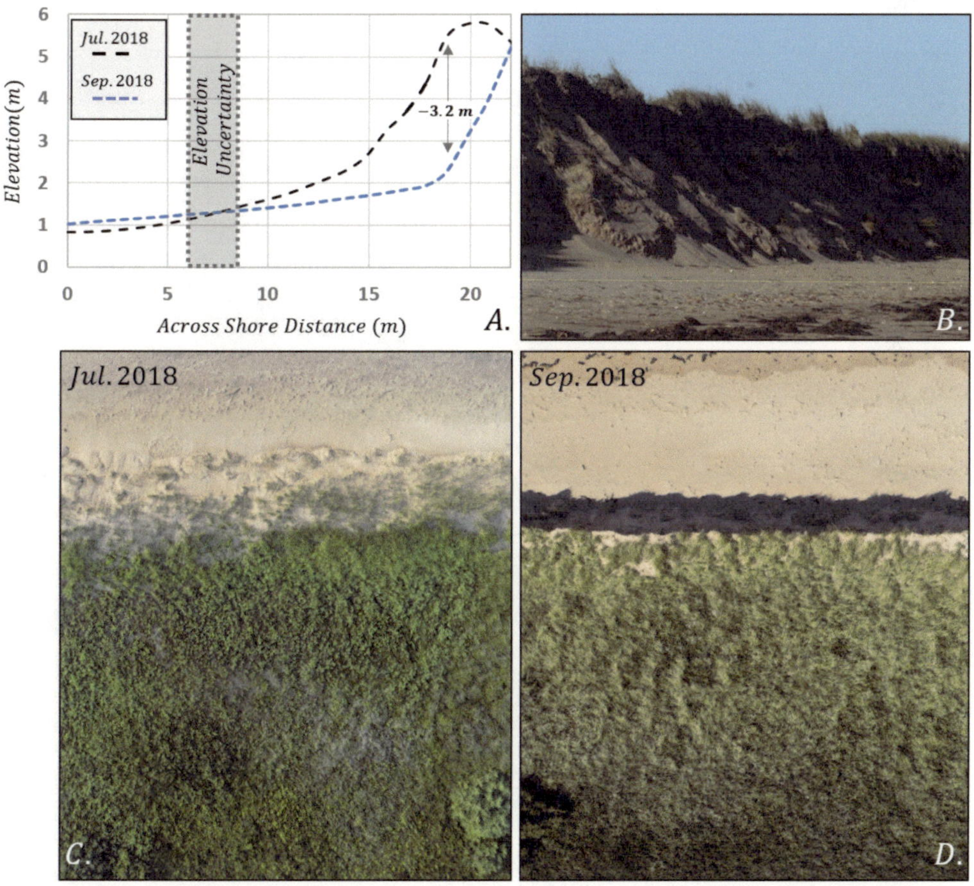


**Figure 7.**
*A topographic change profile sampled from Brackley Beach (BB) east was measured from the pre-storm Dorian (SD) LiDAR DTM and and post-SD UAV DSM (A) BB east experienced significant scarping that eroded the frontal section of the foredune and lead to large volumetric losses (A and B). Aerial UAV images from pre-SD (C) and post-SD (D) show the removal of the stoss slope and formation of a steep scarp at the former crestline. Images from pre-SD (C) and post-SD (D).*

significant downcutting and volumetric loss (**Figure 7A** and **B**). Only a small proportion of this eroded sediment was deposited on the upper beach surface suggesting that the majority of the sediment was moved further seaward, beyond the detectable range of our SfM DSM. Pre-SD, UAV images from July 2019 (**Figure 7C**) show a sparsely vegetated stoss slope extending ~10 m seaward of the dune crest. Post-SD, UAV images from September 2019 show the stoss slope has been entirely eroded leaving behind a steep scarp face formed at the crest line (**Figure 7B** and **D**). The foredunes in this area consistently displayed stoss slopes exceeding the angle of repose for dry sand ~>34° and appeared to be temporarily maintained by surface moisture post-SD. Additional slumping of the surface is expected and may lead to further instability and reduction of dune height at the eastern extent of BB East.

The mid and western sections of BB recorded a continuous ~1–2 m scarp at the base of the stoss slope, resulting in lower volumetric losses from the foredune in these areas (**Figure 8A** and **B**). At BB West, the transition between erosion and deposition occurs much closer to the dune toe, compared to BB east. Also, sediment deposition in this section of BB (**Figure 8A** and **B**) contained a high proportion of the volume eroded from the foredune (**Figure 8A**). In July 2019, prior to SD, the stoss slope of the foredune typically extended ~13 m or more seaward of the crest (**Figure 8C**). Sediment accumulation was aided by the increased seaward extent of

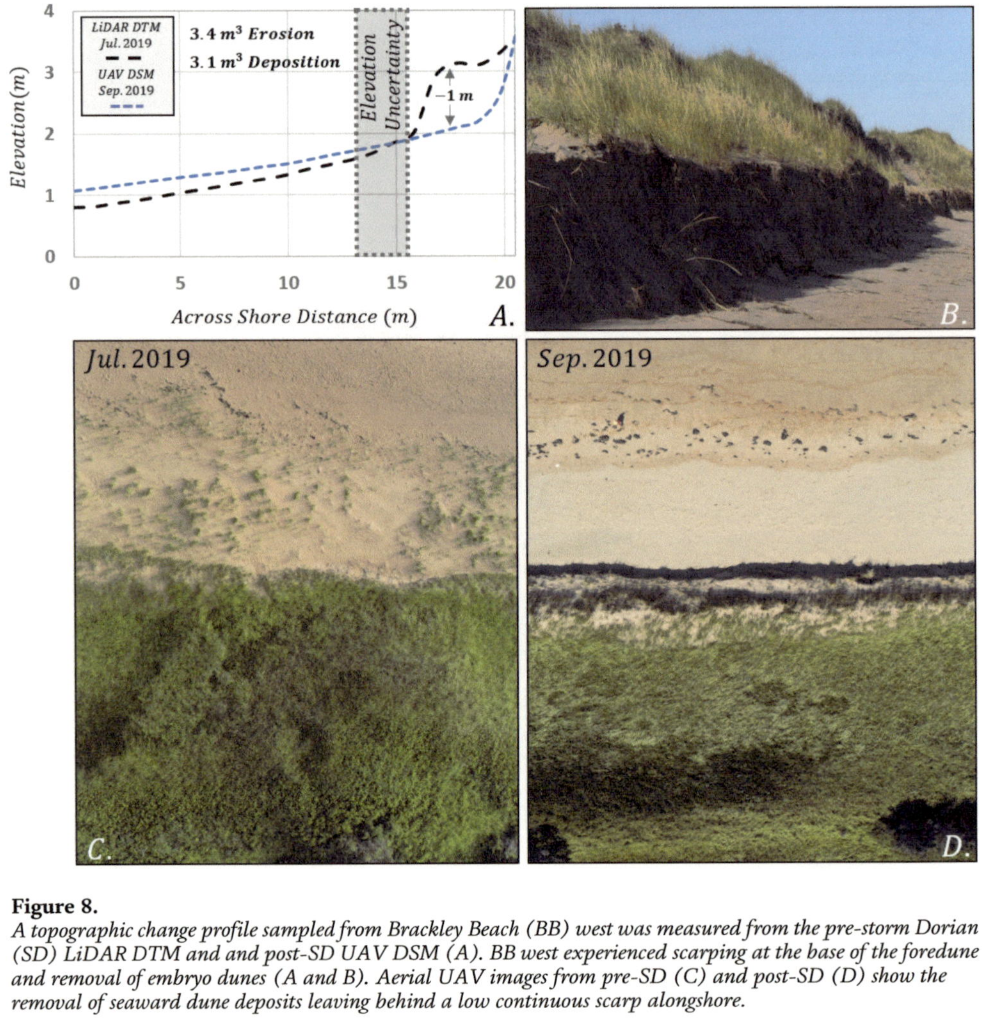


**Figure 8.**
*A topographic change profile sampled from Brackley Beach (BB) west was measured from the pre-storm Dorian (SD) LiDAR DTM and and post-SD UAV DSM (A). BB west experienced scarping at the base of the foredune and removal of embryo dunes (A and B). Aerial UAV images from pre-SD (C) and post-SD (D) show the removal of seaward dune deposits leaving behind a low continuous scarp alongshore.*

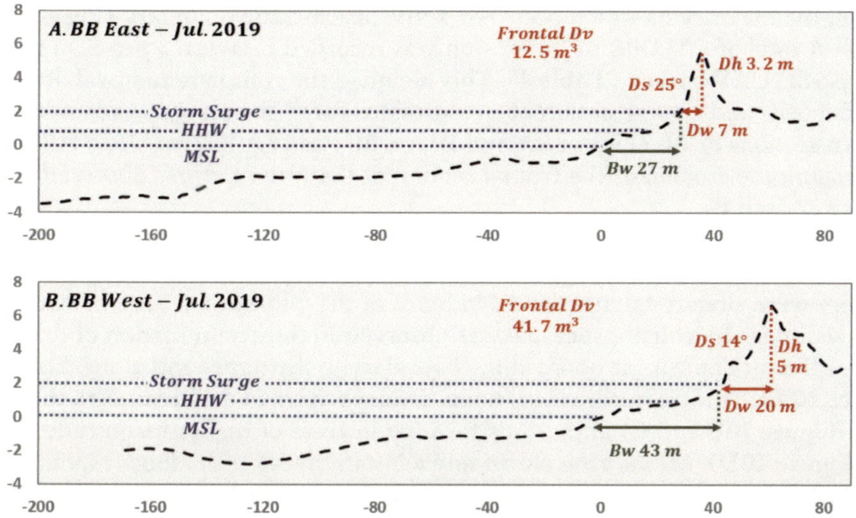


**Figure 9.**
*Alongshore variability of antecedent (i.e., prior to the storm event) morphometry, measured form the LiDAR bathymetric and topographic survey at Brackley Beach (BB; A) east and BB west (B). BB east (A) displayed a higher sloping near shore, narrower beach ($B_w$) and dune width ($D_w$), lower dune height ($D_h$), higher slope ($D_s$), and lower ($D_v$) than BB west (B). These antecedent beach-dune metrics indicate that BB east was more vulnerable to storm erosion that resulted in higher magnitude scarping in this area of BB.*

the vegetation during the summer months (**Figure 8C**) and promoted the intermittent growth of embryo dunes seaward of the established foredune (e.g., **Figure 8A** and **C**). Following SD in September 2019, the scarp line has eroded into the lower stoss slope and has removed any alongshore embryo dunes present pre-SD (**Figure 8D**).

The variability of storm impacts observed at BB were likely controlled by the antecedent morphology that existed prior to SD. From the LiDAR DTM, BB east was observed to have a higher nearshore slope and closer inner bar structure, in comparison to BB west (**Figure 9A** and **B**). This could have led to higher wave energy and erosive potential at BB East during SD, and stronger counter current moving sediment further seaward post-SD (e.g., **Figure 7A** and **8A**). Beach width ($B_w$) at BB East was also significantly narrower than BB West and likely would have increased the exposure of the foredune to wave run up, despite a lower storm surge level relative to the dune toe (**Figure 9A** and **B**). In comparison to BB West, a combination of foredune metrics including narrower dune width ($D_w$), shorter dune heights ($D_h$), higher dune slopes ($D_s$), and lower dune volumes ($D_v$) indicate that BB East was more vulnerable to storm induced erosion (**Figure 9A** and **B**). Foredunes that displayed a narrower $D_w$ and high $D_s$ at BB East experienced the highest magnitude of erosion resulting from major slope failures. While the foredune at BB West appeared to only lose a small proportion of the frontal $D_v$, a significant reduction of $D_v$ at BB East could lead to further lowering of the foredune making this area of BB increasingly vulnerable to future storm events.

## 5. Discussion

UAV monitoring has allowed for the rapid assessment of Brackley Beach (BB), 11 days following Storm Dorian (SD). Projected 7–8 m significant wave heights and 1.2 m storm surge levels associated with SD resulted in a highly erosive post storm surface, the majority of which occurred at the beach-dune boundary (**Figure 6A** and **B**). A total of ~11,000 $m^3$ of erosion was recorded between a pre-SD LiDAR and post-SD UAV survey (**Table 1**). This included the complete removal of embryo dune deposits and the formation of a continuous foredune scarp in the middle and western sections of BB (**Figure 8A** and **B**). At BB East, significant slope failure and high magnitude erosion of the frontal section of the foredune was observed (**Figure 7A** and **B**).

Despite a highly erosive post storm surface, a follow up field investigation in November 2019, or 2 months post-SD, observed the that the initial stages of dune recovery were already taking place. Evidence of the widespread remobilization of beach sediment by eolian processes was observed in the accumulation of dry ripple laden sediment deposits at beach-dune boundary, both across and alongshore (**Figure 10A**). This has resulted in the initial deposition of sediment into the dune scarp (**Figure 10B** and **C**) and ramps forming in areas of higher magnitude deposition (**Figure 10D**). Monitoring storm impacts can reveal initial topographic adjustments resulting from a single event; however, subsequent beach-dune responses can provide a broader understanding on the resiliency of sandy coastlines. Given the low cost, rapidity, and high resolution of UAV SfM surveys, researchers will increasingly have access to high resolution geo-spatial data sets to continuously monitor both short- and longer-term coastal dynamics.

Although UAV SfM research has increased significantly over the last few years, studies that have reported topographic or volumetric change from beach-dune systems are still limited [37, 38, 40, 44, 45]. In part, this has been due to the inherent difficulties in data collection, post-processing, and handling of survey

**Figure 10.**
*Initial dune recovery was observed during a follow up field site assessment in November 2019, or 2 months following storm Dorian (SD). Alonshore, dry sediment has begun to accumulate at the beach-dune boundary (A). This has resulted in the initial accumulation of sediment at the dune scarp (B and C) and ramp development in areas of higher magnitude deposition (D).*

artifacts (e.g., vegetation). This chapter has attempted to address some of these common challenges using basic quality control measures and handling of data uncertainties. In doing so, we were able to confidently quantify alongshore topographic and volumetric changes resulting from a storm event. Survey uncertainties in our post-storm measurements were mitigated due the implementation of quality controls to produce high accuracy (i.e., low RMSE) surface models from UAV surveys. Furthermore, environmental uncertainties were reduced in large part by the wave induced removal of vegetation from the beach dune boundary post-SD. While vegetation uncertainty did not significantly constrain our post storm measurements of the dune scarp, subsequent dune recovery studies will need to develop new strategies to handle uncertainty introduced by vegetation recolonization of the backshore zone.

The quality controls measures, discussed in Section 3, partially include data processing techniques that are unique to the Pix4D software. It is currently beyond the scope of this chapter and expertise of the authors to provide a comprehensive review of different SfM software. Therefore, the remainder of this discussion will focus on addressing survey and environmental uncertainties.

### 5.1 Survey and environmental uncertainties

UAV SfM surveys have proven to be highly accurate with RMSE values, typically <10 cm [36, 38–40, 43]. This level of accuracy is capable of monitoring topographic adjustments associated with storm impacts. However, the uncertainty of topographic change measurements should be reported especially when changes approach the same magnitude of the survey error. While the RMSE values are commonly reported, interpretation of how these values influence uncertainty in topographic change measurements are not. To address survey error, a previous study used RMSE as a threshold to monitor topographic change [44]. In this study, the RMSE value was 16 cm and only positive or negative values in exceedance of $\pm$16 cm were reported. This method is effective at filtering data that could represent no change, but there still could be a high level of uncertainty associated with low magnitude changes. For instance, 17 cm of deposition recorded from a 1 m $\times$ 1 m pixel would result in a volume change ($\Delta V$) and volume change uncertainty ($\Delta VU$) 0.17 $m^3 \pm 0.16$ $m^3$, respectively. Or in other words, 94% of the $\Delta V$ would be within the range of uncertainty. This indicates that interpreting the significance of low magnitude topographic changes should be done carefully and only after survey uncertainty has been reported and adequately addressed.

Results presented in this chapter limited survey uncertainty by applying a topographic change threshold ($CL_{95\%}$) that approximates to a minimum detectable range of 1.96 times the propagated RMSE value. Applying $CL_{95\%}$ effectively increased the confidence in $\Delta V$ measurements by filtering out small magnitude changes that are disproportionately responsible for high levels of $\Delta VU$. For example, 42% of the total observable area of post-SD BB was classified as a zone of elevational uncertainty but, once removed, this only resulted in a change of total $\Delta V$ reported by 3% (**Table 1**). Furthermore, the ratio between $\Delta V$ and $\Delta VU$ dropped from 49 to 23% prior to- and after the$CL_{95\%}$ threshold was applied. It is important to note that $\Delta V$ below the $CL_{95\%}$ threshold, or within the zone of uncertainty, may represent actual change; however, to increase confidence in reported $\Delta V$ values it is important to systematically address survey uncertainty. This chapter has demonstrated that thresholding can be an effective approach to reduce topographic change uncertainties, but the most effective approach remains the quality of repeatable survey strategies and data processing (e.g., discussed in Section 3). For coastal studies, maintaining a high level of survey accuracy limits the areal extent and vertical range of elevational uncertainty and will allow for low magnitude and morphologically significant topographic adjustments (e.g., dune recovery) to be observed through time.

Environmental uncertainties within UAV SfM data also pose a significant challenge as they are space and time dependent. Images taken at the water boundary often captures breaking waves, wave run up, or low contrast moist surfaces in the nearshore and foreshore zones. This can lead to insufficiencies in ATP generation during SfM processing [36, 37, 41] and inconsistencies on the seaward extent in which repeat topographic change measurements are taken. Previously, mean sea level (MSL; [45]) and the wet-dry line [44] have been identified as the observable extent of coastal DSMs. These provide a reasonable estimate of the average annual water line or the water line at a particular time but can become inconsistent when monitoring spatially and/or temporally extensive surveys.

For instance, 2.5 km of post-SD BB was surveyed over a two-day period with approximately 4 h of surveying time per day. Considering the 6 h semi-diurnal (i.e., two high and low) tidal cycle at BB, variability of the across shore extent of the wet-dry line was captured during our alongshore survey. Beyond daily cycles, monthly and annual cycles effect the magnitude of the tidal range and can result in significant variability of the water line boundary during repeat surveys. This may introduce uncertainty when quantifying multi-temporal sediment budgets (e.g., $\Delta V$) and beach metrics (e.g., beach width) as these values may be over- or under-estimated depending on unique spatiotemporal tidal patterns captured while surveying. Alternatively, this chapter used the HHW line as consistent elevation that is typically above the intertidal foreshore zone regardless of tide cycles and ranges. While this approach is conservative in constricting the seaward survey extent, it provides a repeatable methodology to limit environmental uncertainty at the water line boundary.

At the beach-dune boundary, perhaps the biggest challenge in producing repeat UAV surveys in sandy coastal systems is the presence and handling of vegetation. Vegetation can directly lead to over-estimation of surface elevation values in SfM DSMs or point clouds [41]. Furthermore, any topographic change values that have not removed vegetation would be inaccurate as they could represent either a change in vegetation density or canopy height at the seasonal scale or loss of vegetation following a disturbance [38]. Post-SD, elevated storm surge removed vegetation seaward of the scarp along BB and did not have a significant influence our ability to measure topographic change; however, following storm events or seasons, vegetation recolonization could constrain subsequent monitoring surveys during periods of high growth rates.

A UAV vegetation monitoring study at BB from July 2019 sampled vegetation density, almost entirely low profile (i.e., ~50–60 cm) *Ammophila breviligulata* (*Ab*), both across- and alongshore during the peak growing season [53]. Across shore (i.e., from the shoreline to the back dune zone), vegetation density tended to fluctuate between the vegetation line and dune toe, followed by a rapid increase in density up the stoss slope, before reaching 100% density near the crest (**Figure 11A**). Alongshore, the overall density between the shoreline and dune crest, sampled at every 10 m along BB, ranged from 10 to 60% (**Figure 11B**). Taking into account uncertainty introduced with the fluctuating water line, only a narrow band landward of the foreshore and seaward of the vegetation line would consistently be observable without error introduced by spatially and seasonally variant vegetation density patterns.

A high resolution SfM derived point cloud of ~2000 points per $m^2$ was produced from a preliminary UAV survey during July 2019 (**Figure 11B**). The high vegetation density during the peak growing season is evident in the backshore; however, a large number of 'bare earth' points are also visible (**Figure 11B**). An attempt to filter

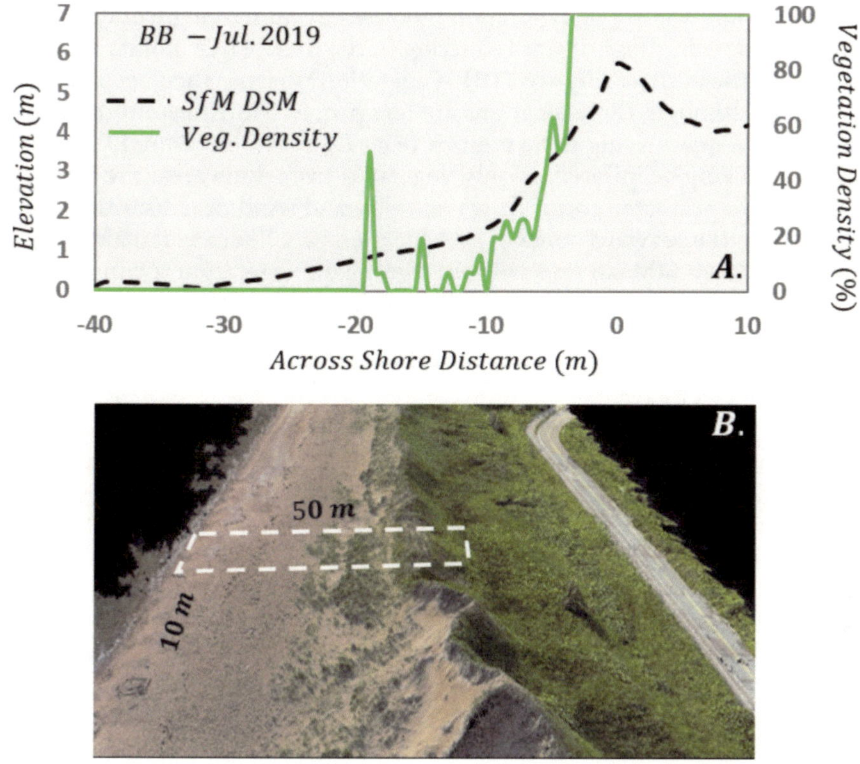


**Figure 11.**
*An across profile and vegetation density, measured from aerial UAV photos taken at Brackley Beach (BB) in July 2019, show variable density at the seaward extent followed by rapid increase in density from the mid stoss slope to dune crest (A). The UAV SfM point cloud generated for this same section of BB, show variability in bare sand and vegetated points captured in the point cloud as you move landward from the sparsely vegetated beach-dune boundary to highly vegetated back dune zone (B).*

the point cloud was completed in Pix4D and consisted of classifying and removing supra-surficial features and produce a 'bare earth' DTM. The standard Pix4D vegetation class, 'high vegetation', may be suitable for other prominent vegetation types (e.g., trees) but was not designed for and, therefore, unable to adequately classify and remove low profile vegetation that typically grows on the backshore (e.g., *Ab*; **Figure 11B**).

Vegetation filtering can be improved by applying algorithms that have been specifically designed for removing beach grasses. For example, a recent study [41] applied a vegetation filtering method originally designed to filter *Ab* from a TLS derived point cloud [51]. Results of this study show that sparsely vegetated foredunes can be effectively filtered from relatively low density SfM derived point clouds (i.e., ~100 points per $m^2$), but do not perform as well in areas of higher vegetation densities, relative to higher density point clouds (i.e., ~2000 points per $m^2$) produced from a TLS sensor. Vegetation is often intermittent at the beach-dune boundary, suggesting that point cloud filtering approaches, especially when applied to higher density SfM point clouds (e.g., **Figure 11B**), could accurately monitor low magnitude topographic changes associated dune building and recovery. In this regard, further studies are needed to test the capability of different vegetation filtering algorithms to remove variant patterns of backshore vegetation growth. Handling of vegetation remains problematic and, without using a suitable filtering method or quantifying the additional uncertainty that it may introduce, should not be included in topographic change measurements.

The density and variety of vegetation increases at BB moving from the crest landward. As a result, a significant reduction in the 'bare earth' points become problematic in these areas (**Figure 11B**). Vegetation filtering then becomes a less viable option. Although, these locations are less prone to low magnitude topographic changes, quantifying dune metrics (e.g., $D_h$ or $D_v$) measured from UAV surveys would likely be inflated. Highly vegetated back dune areas are typically stable as they are protected from the wave, tidal, and wind processes that are actively shaping the seaward zones. These locations are likely to remain a 'zone of uncertainty' in UAV SfM surveys without significant vegetation removal or burial through blowout development, breaching, or overwash events.

Therefore, current UAV SfM applications for monitoring repeat topographic changes occurring in coastal systems are likely to be constrained to the backshore. The accuracy of UAV SfM data on unvegetated surfaces has been demonstrated and there is significant potential to increase the accuracy of sparsely vegetated foredune slopes by applying filtering algorithms (e.g., [41]). It is evident that challenges still remain in resolving water and vegetation boundaries; however, the benefits of UAVs are also clear as they provide an accessible cost-effective method to produce high resolution and spatially extensive surveys of sandy coastal systems. The number of UAV SfM monitoring studies in coastal systems are likely to increase in the coming years, but in order for these studies to confidently report topographic adjustments between the beach-dune boundary, addressing data uncertainties and improvements in vegetation filtering methods are needed.

## 6. Conclusion

This chapter has addressed common problems associated with UAV SfM research on sandy coastlines by presenting a methodology for survey and quality control measures, handling of uncertainties, and the interpretation of storm impacts. This introduction to basic geo-spatial techniques and considerations is aimed at coastal researchers who are developing UAV monitoring strategies. UAVs are becoming increasingly used to monitor beach-dune dynamics, so a systematic approach to address these issues are needed. This study has shown the ability of UAV SfM to accurately report the topographic adjustments of a sandy coastline that has been impacted by a storm event. However, it is also noted that our ability to confidently report these changes was aided in the removal of vegetation at the beach dune boundary. Post-storm recovery of the beach-dune system will be coincident with periods of vegetation growth and, thereby additional environmental uncertainties will be introduced.

Future studies can advance upon these current methodological considerations, particularly regarding the application of vegetation filtering algorithms to reduce environmental uncertainties and constraints. A review of current filtering techniques applied to UAV SfM point clouds, and specifically aimed at removing vegetation characteristic of the backshore, could determine to what extent these systems are capable of continuously monitoring topographic changes occurring at the beach-dune boundary. Using UAV SfM systems to accurately monitor subsequent foredune recovery are dependent on addressing the spatiotemporal uncertainties discussed in this chapter and resolving remaining limitations in handling backshore vegetation. As UAV SfM studies continue to increase in volume, it is critical that uncertainties are addressed in order to confidently monitor topographic adjustments resulting from storm impacts, post-storm recovery, and the longer-term resiliency of beach-dune systems.

## Acknowledgements

The authors would like to thank Prince Edward Island National Park for permitting our field work at Brackley Beach, logistical assistance, and supplying of equipment. We would also like to thank CBCL Limited for supplying the baseline LiDAR survey used in this chapter. Finally, a special thank you to Phillipe Wernette and Jacob Lehner for their invaluable field assistance.

## Author details

Alex Smith*, Brianna Lunardi, Elizabeth George and Chris Houser
School of the Environment, University of Windsor, Windsor, Ontario, Canada

*Address all correspondence to: absmith9@uwindsor.ca

## References

[1] Sherman DJ, Bauer BO. Dynamics of beach-dune systems. Progress in Physical Geography. 1993;**17**(4):413-447

[2] Shepard FP. Beach Cycles in Southern California. Washington, DC: Beach Erosion Board, Corps of Engineers; 1950

[3] Bascom WN. Characteristics of natural beaches. Coastal Engineering Proceedings. 1953;**4**:10

[4] Houser C. Synchronization of transport and supply in beach-dune interaction. Progress in Physical Geography. 2009;**33**(6):733-746

[5] Cowell PJ, Thom BG. Morphodynamics of Coastal Evolution. Cambridge, United Kingdom and New York, NY, USA: Cambridge University Press; 1994

[6] Houser C, Ellis J. Beach and dune interaction. Treatise on Geomorphology. 2013;**10**:267-288

[7] Wernette P, Houser C, Weymer BA, Everett ME, Bishop MP, Reece B. Influence of a spatially complex framework geology on barrier island geomorphology. Marine Geology. 2018; **398**:151-162

[8] Masselink G, Castelle B, Scott T, Dodet G, Suanez S, Jackson D, et al. Extreme wave activity during 2013/ 2014 winter and morphological impacts along the Atlantic coast of Europe. Geophysical Research Letters. 2016; **43**(5):2135-2143

[9] Davidson-Arnott RG. Conceptual model of the effects of sea level rise on sandy coasts. Journal of Coastal Research. 2005:**21**(6):1166-1172

[10] Houser C, Hamilton S. Sensitivity of post-hurricane beach and dune recovery to event frequency. Earth Surface Processes and Landforms. 2009;**34**(5): 613-628

[11] Kandrot S, Farrell E, Devoy R. The morphological response of foredunes at a breached barrier system to winter 2013/2014 storms on the southwest coast of Ireland. Earth Surface Processes and Landforms. 2016;**41**(14):2123-2136

[12] Houser C, Hapke C, Hamilton S. Controls on coastal dune morphology, shoreline erosion and barrier island response to extreme storms. Geomorphology. 2008;**100**(3–4):223-240

[13] Jewell M, Houser C, Trimble S. Initiation and evolution of blowouts within Padre Island National Seashore, Texas. Ocean & Coastal Management. 2014;**95**:156-164

[14] Jewell M, Houser C, Trimble S. Phases of blowout initiation and stabilization on Padre Island revealed through ground-penetrating radar and remotely sensed imagery. Physical Geography. 2017;**38**(6):556-577

[15] Hesp P. Foredunes and blowouts: Initiation, geomorphology and dynamics. Geomorphology. 2002;**48** (1–3):245-268

[16] Houser C, Wernette P, Weymer BA. Scale-dependent behavior of the foredune: Implications for barrier island response to storms and sea-level rise. Geomorphology. 2018;**303**:362-374

[17] Houser C, Wernette P, Rentschlar E, Jones H, Hammond B, Trimble S. Post-storm beach and dune recovery: Implications for barrier island resilience. Geomorphology. 2015;**234**:54-63

[18] Houser C, Barrineau P, Hammond B, Saari B, Rentschler E, Trimble S, et al. Role of the foredune in controlling barrier island response to sea level rise. In: Barrier Dynamics and

Response to Changing Climate. Cham: Springer; 2018. pp. 175-207

[19] Short AD, Hesp PA. Wave, beach and dune interactions in southeastern Australia. Marine Geology. 1982;**48** (3–4):259-284

[20] Bagnold RA. The nature of saltation and of 'bed-load'transport in water. Proceedings of the Royal Society of London. A. Mathematical and Physical Sciences. 1973;**332**(1591):473-504

[21] Namikas SL, Edwards BL, Bitton MC, Booth JL, Zhu Y. Temporal and spatial variabilities in the surface moisture content of a fine-grained beach. Geomorphology. 2010;**114**(3):303-310

[22] Hesp PA. A review of biological and geomorphological processes involved in the initiation and development of incipient foredunes. Proceedings of the Royal Society of Edinburgh, Section B: Biological Sciences. 1989;**96**:181-201

[23] Suanez S, Cariolet JM, Cancouët R, Ardhuin F, Delacourt C. Dune recovery after storm erosion on a high-energy beach: Vougot Beach, Brittany (France). Geomorphology. 2012;**139**:16-33

[24] Sallenger AH Jr. Storm impact scale for barrier islands. Journal of Coastal Research. 2000;**1**:890-895

[25] Stockdon HF, Sallenger AH Jr, Holman RA, Howd PA. A simple model for the spatially-variable coastal response to hurricanes. Marine Geology. 2007;**238**(1–4):1-20

[26] Houser C, Mathew S. Alongshore variation in foredune height in response to transport potential and sediment supply: South Padre Island, Texas. Geomorphology. 2011;**125**(1):62-72

[27] Houser C. Alongshore variation in the morphology of coastal dunes: Implications for storm response. Geomorphology. 2013;**199**:48-61

[28] Dolan R, Hayden B. Storms and shoreline configuration. Journal of Sedimentary Research. 1981;**51**(3): 737-744

[29] Suter JR, Nummedal D, Maynard AK, Kemp P. A process-response model for hurricane washovers. In: Coastal Engineering. 1982. pp. 1459-1478

[30] Orford JD, Carter RWG. Mechanisms to account for the longshore spacing of overwash throats on a coarse clastic barrier in Southeast Ireland. Marine Geology. 1984;**56**(1–4):207-226

[31] Weymer BA, Houser C, Giardino JR. Poststorm evolution of beach-dune morphology: Padre island national seashore, Texas. Journal of Coastal Research. 2015;**31**(3):634-644

[32] Lorenzo-Trueba J, Ashton AD. Rollover, drowning, and discontinuous retreat: Distinct modes of barrier response to sea-level rise arising from a simple morphodynamic model. Journal of Geophysical Research: Earth Surface. 2014;**119**(4):779-801

[33] Cook KL. An evaluation of the effectiveness of low-cost UAVs and structure from motion for geomorphic change detection. Geomorphology. 2017;**278**:195-208

[34] Snavely KN. Scene reconstruction and visualization from internet photo collections [doctoral dissertation, PhD thesis]. Seattle, Washington, USA: University of Washington; 2008

[35] Westoby MJ, Brasington J, Glasser NF, Hambrey MJ, Reynolds JM. 'Structure-from-Motion' photogrammetry: A low-cost, effective tool for geoscience applications. Geomorphology. 2012;**179**:300-314

[36] Gonçalves JA, Henriques R. UAV photogrammetry for topographic monitoring of coastal areas. ISPRS

Journal of Photogrammetry and Remote Sensing. 2015;**104**:101-111

[37] Turner IL, Harley MD, Drummond CD. UAVs for coastal surveying. Coastal Engineering. 2016; **114**:19-24

[38] Scarelli FM, Sistilli F, Fabbri S, Cantelli L, Barboza EG, Gabbianelli G. Seasonal dune and beach monitoring using photogrammetry from UAV surveys to apply in the ICZM on the Ravenna coast (Emilia-Romagna, Italy). Remote Sensing Applications: Society and Environment. 2017;**7**:27-39

[39] Mancini F, Dubbini M, Gattelli M, Stecchi F, Fabbri S, Gabbianelli G. Using unmanned aerial vehicles (UAV) for high-resolution reconstruction of topography: The structure from motion approach on coastal environments. Remote Sensing. 2013;**5**(12):6880-6898

[40] Brunier G, Fleury J, Anthony EJ, Pothin V, Vella C, Dussouillez P, et al. Structure-from-Motion photogrammetry for high-resolution coastal and fluvial geomorphic surveys. Géomorphologie: relief, processus, environnement. 2016;**22**(2):147-161

[41] Guisado-Pintado E, Jackson DW, Rogers D. 3D mapping efficacy of a drone and terrestrial laser scanner over a temperate beach-dune zone. Geomorphology. 2019;**328**:157-172

[42] Laporte-Fauret Q, Marieu V, Castelle B, Michalet R, Bujan S, Rosebery D. Low-cost UAV for high-resolution and large-scale coastal dune change monitoring using photogrammetry. Journal of Marine Science and Engineering. 2019;**7**(3):63

[43] Papakonstantinou A, Topouzelis K, Pavlogeorgatos G. Coastline zones identification and 3D coastal mapping using UAV spatial data. ISPRS International Journal of Geo-Information. 2016;**5**(6):75

[44] Casella E, Rovere A, Pedroncini A, Stark CP, Casella M, Ferrari M, et al. Drones as tools for monitoring beach topography changes in the Ligurian Sea (NW Mediterranean). Geo-Marine Letters. 2016;**36**(2):151-163

[45] Guillot B, Castelle B, Marieu V, Bujan S, Rosebery D. UAV monitoring of 3-year Foredune Partial Recovery from a Severe Winter: Truc Vert Beach, SW France. Journal of Coastal Research. 2018;**85**(sp1):276-280

[46] Suo C, McGovern E, Gilmer A. Coastal dune vegetation mapping using a multispectral sensor mounted on an UAS. Remote Sensing. 2019;**11**(15):1814

[47] Sofonia J, Phinn S, Roelfsema C, Kendoul F. Observing geomorphological change on an evolving coastal sand dune using SLAM-based UAV LiDAR. Remote Sensing in Earth Systems Sciences. 2019;**2**(4):273-291

[48] Dalhousie Coastal Ocean Forecast System–DalCoast-HFX. Canada: Department of Oceanography, Dalhousie University. Available from: http://extrememarine.ocean.dal.ca/dalcoast/web/home.php

[49] Davies M. Geomorphic shoreline classification of Prince Edward Island. Report prepared by Coldwater Consulting Ltd. for the Atlantic Climate Adaptation Solutions Association (ACASA); 2011. p. 66

[50] Staley DM, Wasklewicz TA, Kean JW. Characterizing the primary material sources and dominant erosional processes for post-fire debris-flow initiation in a headwater basin using multi-temporal terrestrial laser scanning data. Geomorphology. 2014;**214**:324-338

[51] Smith A, Gares PA, Wasklewicz T, Hesp PA, Walker IJ. Three years of morphologic changes at a bowl blowout, Cape Cod, USA. Geomorphology. 2017; **295**:452-466

[52] Wheaton JM, Brasington J, Darby SE, Sear DA. Accounting for uncertainty in DEMs from repeat topographic surveys: Improved sediment budgets. Earth Surface Processes and Landforms. 2010;**35**(2): 136-156

[53] Lunardi B, Lehner JD, Houser C, Wernette PA, George E. Alongshore coupling of eco-geomorphological variables of a beach-dune system. In: AGU Fall Meeting, 9 AGU. 2019

Chapter 5

# Recent Advances in Coastal Survey Techniques: From GNSS to LiDAR and Digital Photogrammetry - Examples on the Northern Coast of France

*Olivier Cohen and Arnaud Héquette*

## Abstract

The aim of this chapter is to document the evolution of surveying techniques used for monitoring changes in morphology of beaches and coastal dunes in northern France, beginning with GNSS surveys in the 1990s, followed by airborne topographic LiDAR surveys since 2008, and high-resolution digital photogrammetry data collected by an UAV during recent years. Digital Terrain Models (DTMs) derived from the data obtained from the different techniques were used for monitoring coastal changes, including shoreline evolution, and for computing sediment volume changes across the foreshore and the coastal dunes at different time and spatial scales. A comparison of the results obtained using these different techniques and of their accuracy will be carried out to assess the pro and cons of each surveying technique.

**Keywords:** GNSS, LiDAR, UAV, topographic survey, digital elevation models, coastal geomorphology, northern coast of France, shoreline evolution, sediment budget

## 1. Introduction: from a classical naturalistic approach to a contemporary quantitative approach

Research in coastal geomorphology aims at describing the shapes of coastal landforms (dunes, beaches, cliffs, estuaries, deltas, shorefaces, etc.) at different temporal scales (long to short term, several thousand years to a few hours) and spatial scales (over large to small areas) and understanding the processes of their evolution [1, 2]. This research is frequently applied to the study of risks, for example, in case of shoreline erosion, the evolution of the coastline is mapped, and sediment budgets are calculated to define the hazards threatening human infrastructures.

Many methods are used to detect shoreline changes. They can be classified into three categories depending on the date and duration of the study period and the type of tools used [3].

To map the medium-term coastline evolution from secular to multi-decadal scale and in two dimensions (in plan), diachronic analysis of historic maps, aerial photographs since the mid-twentieth century, and more recent high spatial resolution satellite images are often used. On historic maps that are often imprecise in their drawing, it usually results in approximate measurements. However, they represent valuable documents for a qualitative or naturalistic geo-historic analysis of landscape [3, 4]. On these maps published since the end of the nineteenth century and on vertical aerial photographs available since the 1930s, the successive positions of the coastline can be identified, digitized, and then compared in a Geographical Information Systems (e.g., [5–7]) in order to determine erosion or accretion zones.

Within the second category of methods, in the shorter term, over the last 30 years or so, measurements have been carried out in two ways with high-tech instruments. First, geodetic instruments (e.g., total electronic stations and GNSS) are used to carry out data collection in the field. These measurements are time-consuming and difficult to carry out over large areas; only a fairly limited number of points can be acquired (a few hundred to a few thousand); therefore, only beach profiles (cross section of the range from a few tens to hundreds of meters long) or Digital Terrain Models (DTM) of limited size (a few thousand square meters) and low spatial resolution can be surveyed. Second, also in the short term, telemetry instruments (e.g., ground or airborne LiDAR) that enable the rapid collection of a large amount of elevation data that are transformed into DTM are used.

The third category encompasses the methods of photogrammetric processing that allow to extract elevation data from aerial photographs; this technique also allows to collect a very large amount of data that are transformed into Digital Surface Model (DSM).

The evolution of these techniques can also be conceptualized in a "measurement method paradigm" (**Figure 1**): with the methods of the first category, we start from the images to make measurements. Paper images (maps or aerial photographs) available in paper format are scanned or acquired directly in digital format; geometric deformations are corrected, and they are georeferenced; measurements are then carried out and statistically processed. It should be noted here that the analysis is in two dimensions (in plan). With the methods of the second category, measurements are performed and transformed into images (DTM and DSM), which are the result of calculations and not a direct capture of the reality of the terrain. One of the major interests of these methods is to be able to work in three dimensions: the measurements made are in plan and in altitude. The third category closes the paradigm: the work process starts with images (aerial photographs), continues with measurements, and returns to images (DSM). Here, the analysis is also in three dimensions.

This paradigm has undergone a twofold evolution over the last 25 years or so. First, the transition from 2D to 3D measurements has been a determining factor in coastal geomorphology. This has made it possible to objectify the position of the coastline, which is not always easy to detect in 2D aerial photographs [8]. This also made it possible to calculate volumetric and not just planimetric changes in beaches and dunes. In addition, with LiDAR technology and photogrammetry, there has been a progress toward higher density and accuracy of measurement and results.

The first part of this chapter is devoted to a technical and conceptual synthesis. A bibliometric analysis shows the recent, rapid, and phased success of the use of GNSS, LiDAR, and then airborne photogrammetry in the field of coastal geomorphology. We will briefly describe the techniques, emphasizing their complementarities and their spatial and temporal scales of application. The second part of this chapter is an illustration of the first by selected examples of results obtained on the coast of northern France. The conclusion proposes a synthesis of the advantages and disadvantages of these techniques and discusses some future prospects.

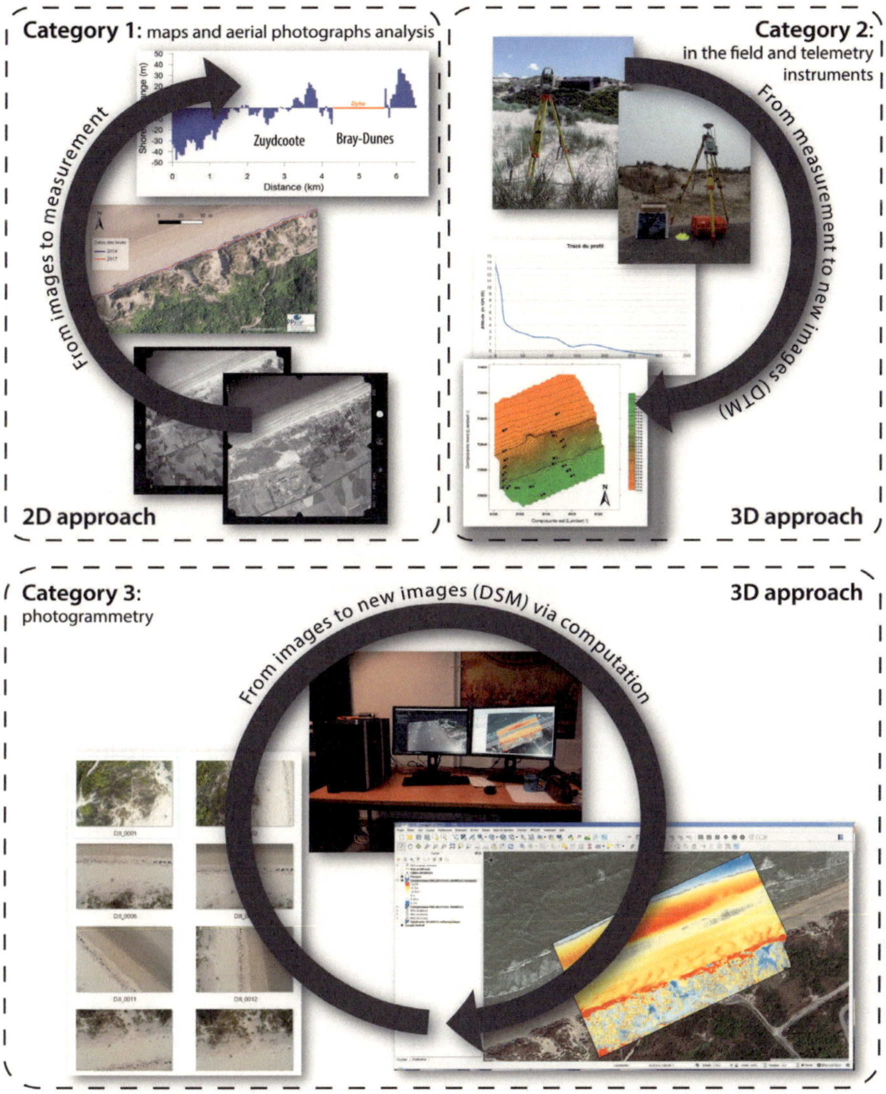


**Figure 1.**
*The measurement method paradigm.*

## 2. From GNSS to LiDAR and photogrammetry: three successful technologies aiming to a higher density of measurements

The use of these three techniques in Earth and environmental sciences began about 30 years ago. They have achieved rapid and phased success as they have been further developed and progressively introduced.

### 2.1 A rapid growth in the use of these techniques

#### *2.1.1 In environmental sciences*

In environmental sciences, the applications of these three techniques are very diverse and operate at different spatial scales.

Differential GNSS is an in-situ measurement system that is used in all areas where an accurate position on the planet's surface is required [9]. With GNSS, for example, time series of measurements can be made at fixed locations, for example, to identify the role of vertical movements of the continent in changes in relative sea level [10] or to map the plate boundary faults [11]. Multiple point measurements are also frequently taken at different locations and interpolated to produce elevation maps, for example, for glacial geomorphological mapping [12].

Light Detection and Ranging (LiDAR) is a telemetry system (remote measurement) that can be used in two main configurations: on the ground with TLS systems and in airborne configuration. As shown in a bibliometric study in the field of earth sciences in Ref. [13], the Terrestrial Laser Scanning (TLS) technique is used to make precise topographic surveys at distance without having to come into contact with the ground, which is very convenient in hard-to-reach terrain such as steep cliffs. Very large numbers of measurements can be made with these instruments, which allow to model the investigated object. Fields of investigation using the techniques are, for example, at fine-scale mineralogy and petrology and at a larger scale structural geology, seismology, volcanology, tsunami hazards, geomorphology, and cryosphere studies [13]. Airborne LiDAR is used to produce accurate elevation maps at a larger scale, usually over large areas. The spectrum of applications in the Earth and ecological sciences is wide [14], for example, for lava flow survey [15] and forestry [16, 17].

Photogrammetry can also be implemented at different scales, from very fine scale, over a few millimeters, for example, to monitor the rock surface weathering [18], to medium scale, over areas ranging from a few thousand square meters to a few hectares, for example, to monitor the evolution of slopes [19, 20] or coastal habitats [21–23]. Finally, it can be deployed on a large scale, over areas of several square kilometers (e.g., geomorphological mapping in high mountain environment) [24].

### *2.1.2 In coastal geomorphology over the past 25 years*

In order to estimate the development of these three techniques in coastal geomorphology in particular, a bibliometric analysis was conducted following a method similar to that of [13] for the use of TLS in Earth sciences and [25] for UAVs in agriculture and forestry.

The Scopus database was used for finding publications (articles and book chapters) in the field of Earth and planetary sciences, from 1995 to 2019, searching for the following keywords: the kind of instrument (gps or dgps or gnss or dgnss/LiDAR/uav or uas or drone), coast or beach or dune or shoreline. Early results often contained intruders (e.g., in the military, maritime navigation, biology, mathematics, or electronic fields). It was therefore necessary to reduce the research panel by excluding keywords and certain journals outside the field of coastal geomorphology.

**Figure 2** shows the results of this analysis for the three types of techniques. It can be seen that GNSS was used earlier than the other two techniques. This is logical since it is the first of the three technologies that have been developed. However, during the last years, the number of mentions of satellite positioning systems decreases. This is probably not because they are less used, rather because they have become commonly used and authors probably no longer feel the need to describe or even mention the technique in the publication. However, these GNSSs are still needed for complementing LiDARs and UAVs.

The LiDAR citation curve reflects the development of this technology during the last decades. In the second half of the 1990s, there are a limited number of publications in which the tool is experimental or is used by only a few organizations that

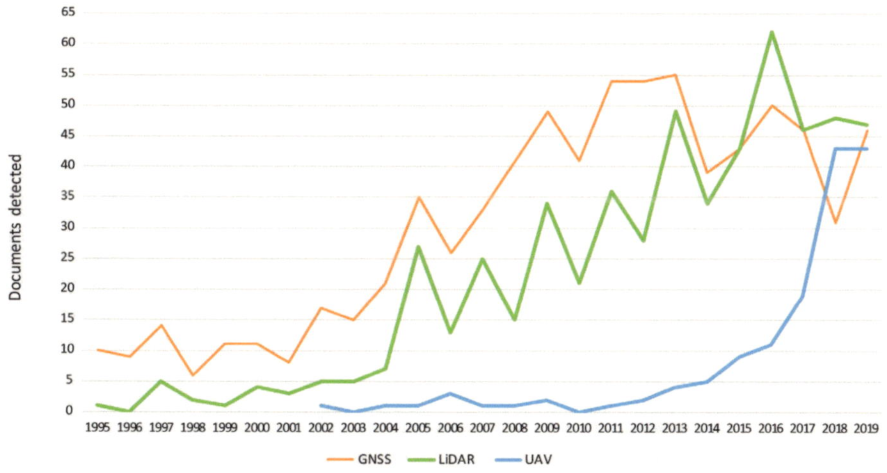


**Figure 2.**
*Results of the bibliometric analysis on GNSS, LiDAR, and UAV mentions from 1995 to 2019.*

have the technical and financial means to use it (e.g., US Army Corps of Engineers USACE) [26]. This technique significantly increased at the beginning of the 2000s whether airborne or TLS (e.g., for monitoring the evolution of cliffs) [27–29]. The use of UAVs did not begin until the early 2000s. This technique took off during the second decade of the twenty-first century with the availability of light aircraft (<2 kg), ready to fly, simple to program and pilot, and financially affordable [22]. UAVs are most often used for photogrammetry (see below) and more rarely to carry small LiDAR systems [30] or multispectral sensors [31]. It is interesting to note that, at the end of the study period, the three techniques are almost equal in terms of citations in the literature: although the use of UAVs started late, it is now as common as that of GNSS and LiDAR.

## 2.2 The use of these techniques in coastal geomorphology

### *2.2.1 Differential GNSS*

Global Navigation Satellite Systems (GNSSs), more commonly known by the acronym GPS, make it possible to measure a position in three dimensions, anywhere on the Earth's surface, using the trilateration principle (with three satellites) [32], that is, by referring to the very precisely known positions of the satellites in their orbits. The use of a radio beacon precisely geo-referenced on land, installed near the study site and communicating with the GNSS receiver via UHF waves, makes it possible to reduce the error margin of measurement, which ranges from several millimeters to centimeters, or even millimeters both horizontally and vertically.

During fieldwork campaigns, when numerous measurements for generating accurate Digital Terrain Model are required, the RTK (Real Time Kinematic) mode is generally used: the GNSS automatically records elevation data points during the operator's movements following a previously defined time step or distance.

Elevation point measurements can be acquired on foot when the areas to be covered are fairly small (a few thousand square meters or even a few hectares) or with an all-terrain vehicle (e.g., a quad where the GNSS is mounted coupled to an inertial station to compensate for the vehicle's movements) driving slowly (between 10 and 20 km/h) for larger and fairly flat areas. Although larger areas can be covered with a vehicle, the density of points collected is rather low. In the example shown in **Figure 3**, the elevation data points were collected using a Leica 1200 GNSS, which

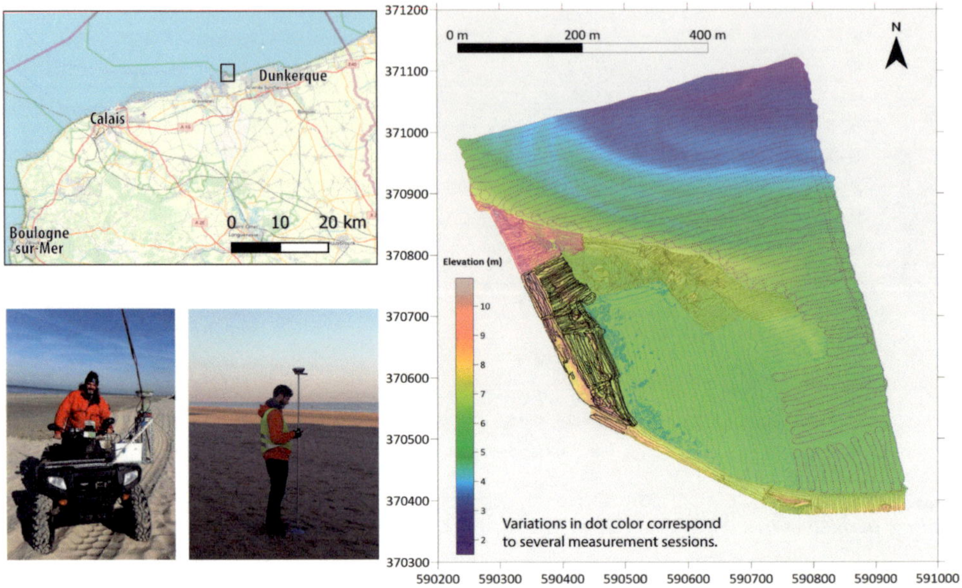


**Figure 3.**
*In the field topographic survey using a GNSS; example of a DTM (the measurement points are overlapped on the DTM).*

accuracy is ±1.5 cm horizontally and ±3 cm vertically, coupled to an inertial station SBG 500-E on a quad. The DTM was calculated by triangulation of Delaunay along the coastline of the port of Dunkirk. The measured points were superimposed on it. The intertidal zone was measured with a GNSS receiver of an all-terrain vehicle; the alignments of the points along the vehicle's trajectory are clearly visible. The upper beach, the foredune, the muddy areas, and the sand and pebble spit were measured on foot because the relief is too constraining for the vehicle; foot surveys enable higher measurement densities. This type of topographic survey is time-consuming: for a survey of 36,360 points, over a surface area of 33.5 ha, resulting in a measurement density of 0.1 point per $m^2$, it took 5 days of work during daytime and low tide periods.

### *2.2.2 Airborne LiDAR*

Airborne LiDAR is an active telemetry instrument. A vector, aircraft [33–35], helicopter, or UAV [30, 36] whose position and altitude are precisely determined using a differential GNNS, carries a side-scanning laser that emits pulses under the vector. The laser beam is emitted toward the ground; the round-trip time of this beam and its "echo" is measured. The speed of propagation of the beam is perfectly known according to the atmospheric conditions (temperature and humidity). For each point aimed at on the ground, several signals are recorded and averaged: this is the distance between the device and the ground surface. Ground control points are checked with GNSS to calibrate the data as accurately as possible.

Airborne LiDAR surveys can provide a large number of measurements over large areas of several square kilometers. The measurement density is higher than that obtained with in situ GNNS. It can typically range from 1 to 2 points/$m^2$, for example, from 1.2 to 1.4 points/$m^2$ after data filtering [37, 38]. In the example shown in **Figure 4**, a Leica HawkEye III topo-bathymetric LiDAR sensor was used with a 500 KHz frequency in infrared spectrum for topographical surveying and a 35 KHz frequency in the green spectrum for bathymetric monitoring in shallow

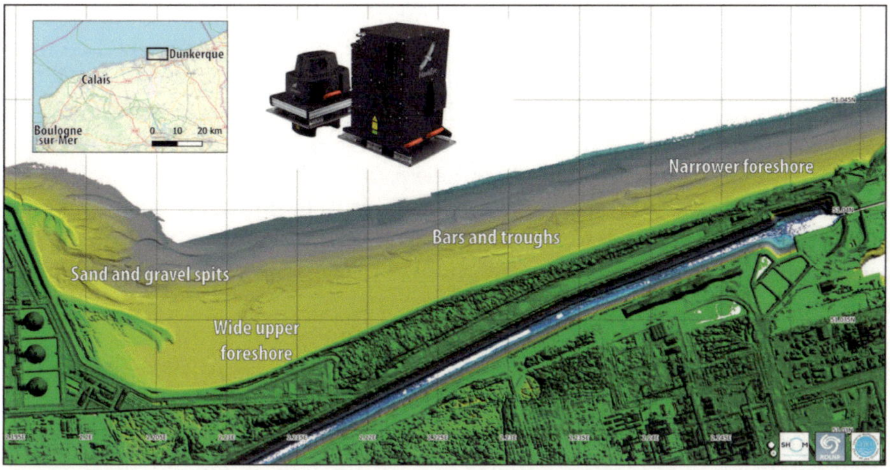


**Figure 4.**
*Extract of an extensive 1 m spatial resolution DTM on the harbor of Dunkirk coast, northern France carried out with a Leica HawkEye III LiDAR (sources: map SHOM, ROLNP on data.shom.fr; picture, Leica Geosystems).*

water. The density of topographical measurements is 12 points/m$^2$ before data filtering. According to the manufacturer, the measurement accuracy is ±20 cm horizontally and ±2 cm vertically. This vertical margin of error is generally noted on bare surfaces (sand and rock) but increases significantly in vegetated areas (25 cm).

Maps produced from LiDAR data are Digital Surface Models (DSM) that show the top of all objects. These DSMs represent the ground relief in bare areas (beach and rocky plains) but may show the tops of vegetation or buildings elsewhere. In geomorphology, however, the object of the study is the relief on the ground and not that on the surface of the vegetation or other objects. To obtain a Digital Terrain Model (DTM), buildings can be removed by using a vector map of the building to delineate the areas of the points to be deleted. At the location of each building, an interpolation is then made with the ground elevations around the perimeter of the building. For vegetated areas, the data are filtered: the first echo of the signal is considered as the one corresponding to the surface of the objects, and the last echo (the furthest from the vector) is assimilated to the ground. DTMs are therefore calculated using the last echoes resulting in lower measurement densities. However, the vegetation is sometimes too dense for the LiDAR signal to reach the ground and for a DTM to be extracted from the DSM; this is notably the case in coastal dunes covered by sea buckthorn (**Figure 4**). DSMs or DTMs are calculated by interpolation from points; they can be represented in vector mode as contour maps or in raster mode where each pixel contains elevation information. With a data point density of 1–2 points/m$^2$, there is no need to aim for a spatial resolution of DSM or DTM finer than 1 m/pixel. **Figure 4** shows the very wide foreshore along the port of Dunkirk, which gradually narrows toward the east. It also shows a morphology of bars and troughs on the foreshore. The sand and gravel spits are clearly visible.

### *2.2.3 Airborne photogrammetry*

Photogrammetry is a nonactive telemetry process because no signal is emitted. It can be defined simply as the "science of making reliable measurements from photographs" [39]. It is most frequently applied to aerial photographs. It allows distance and surface measurements to be made in plan. It is also possible to extract altitude information from two vertical aerial photographs of the same area taken from different angles using the principle of stereoscopy.

Photogrammetry therefore requires a camera: formerly, a silver camera, and nowadays, a high-resolution digital sensor. These cameras are embarked on different types of carriers: airplanes [40], ultralight aircraft [41], or UAV [42–46]. The cameras are installed on stabilized gimbals that compensate for the vector's pitch and roll movements in order to minimize the blurring effects on the photographs.

"Conventional" airborne photogrammetry, using carriers flying at altitudes of 500–5000 m above the ground, provides results with highly variable vertical error margins that largely depend on the spatial resolution of the acquired images and therefore on the flying height. The margin of error can be metric (1.5 m in mountainous areas) [24] or decimetric (30 cm in coastal areas) [39], which is not accurate enough to detect fine detail and small amplitude changes on beaches.

In recent years, airborne photogrammetry by UAV at low altitude (below 150 m) and very high spatial resolution has rapidly gained popularity in coastal geomorphology. As mentioned above, this success has been made possible by the development of digital photogrammetry and the introduction of small, easy-to-use, and affordable civilian UAVs. This success is also due to a technical change in photogrammetry, which was originally only an optical technique, requiring the use of bulky equipment and the permanent intervention of a technician, whereas it is now completely digital and can be automated almost all along the image processing workflow. Specialized software, but relatively easy to use for those who are familiar with geomatics, is available. These programs perform a correspondence analysis of the pixels of the images to "stitch" them together like a panorama; they calculate angles between the shooting points and these pixels in order to calculate depths of field or differences in altitude to represent the relief.

The protocol for organizing a UAV photogrammetry mission is simple. A flight plan is programmed, which forecasts the UAV's trajectory in the form of flight path parallel to each other in the area to be studied. The flight altitude of the aircraft is parameterized, which, depending on the sensor's fineness, determines the spatial resolution of the vertical images: the greater the height, the lower the resolution. For example, with a 20 MPixel sensor, a pixel of 2.41 μm, a focal length of 8.8 mm (DJI Phantom IV Pro or Mavic 2 Pro UAV), a flight height of 50 m, the spatial resolution of an image will be 1.4 cm/pixel.

$$R = H \times 100 \times \left[\frac{P \times 10^{-4}}{F \times 10^{-1}}\right] \tag{1}$$

where $R$ is the spatial resolution of the image in cm/pixel, $H$ is the flying height in meters, $P$ is the sensor pixel size in microns, and $F$ is the sensor focal length in millimeters.

Overlap between two successive photographs and lateral overlap between two parallel strips of photographs are programmed (80 and 60% are recommended, respectively). On the UAV's speed depends on these parameters. The speed must be slow enough to avoid blurring on the photographs. In the example shown in **Figure 5**, the flight height of the DJI Mavic 2 Pro was 65 m, and the speed was 5 m/s. Oblique aerial photographs can also be integrated into the data to be processed: they help to better model steep slope reliefs such as cliffs or foredune fronts. The flight plan is recorded and used for repeated overflights at different dates. During each campaign, ground control points, evenly distributed over the study area, are measured using GNSS. The coordinates of these control points allow the DSM to be set very precisely in plan and in altitude.

Digital photogrammetric processing with very high spatial resolution allows to generate a very high density of measurement points and to calculate very fine DSMs. The DTM, as shown in **Figure 5**, was calculated with an average density of 269 points/m$^2$. The average error margin of the measurements is 3.2 cm in the plane and

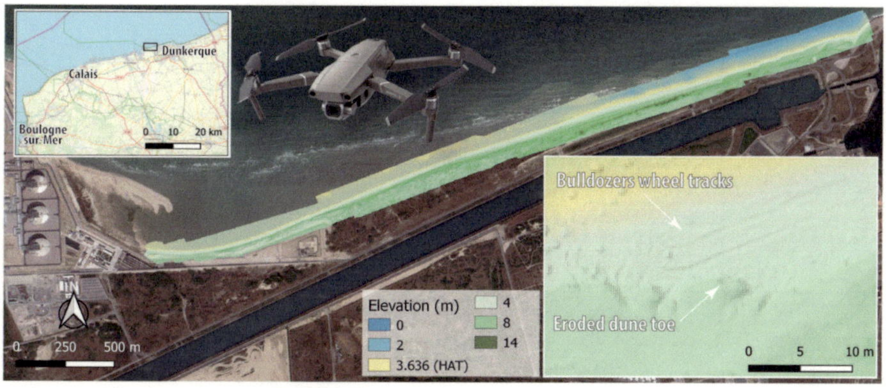


**Figure 5.**
*Example of a very high resolution DSM computed from photogrammetric survey carried out with a DJI Mavic 2 Pro UAV on the harbor of Dunkirk coast, northern France.*

2.5 cm at altitude with a spatial resolution of 6 cm/pixel. The sharpness is such that when zooming in, wheel tracks of bulldozers replenishing the beach a little further east can be seen. Such spatial resolution of the DSM can be achieved, thanks to the very high density of the points. As with LiDAR data, photogrammetric measurements concern the surface of objects: the transformation of DSMs into DTMs is sometimes possible for spatially well-defined objects (buildings and small groves of vegetation) using the same methods. However, such high accuracy DTMs can hardly be obtained in densely vegetated areas. Interpretations of elevation changes must then be very cautious. An apparent increase in altitude may be the consequence of vegetation growth and not of sand accretion (see Section 3 of this chapter).

A major interest of photogrammetry compared to LiDAR is that it allows the acquisition of aerial photographs on which many detailed observations can be distinguished (e.g., sand/vegetation limit on the upper beach and type of vegetation). These photographs, each covering a small area, can be assembled to form a full orthophotograph [39]. They are corrected for optical distortions using precisely known shooting parameters and recorded in the exif metadata of the files; distortions due to relief are corrected using DSM.

### 2.3 Synthesis

GNSS, LiDAR, and photogrammetry can be classified in several ways. A distinction can be made between ground (GNSS and TLS) and airborne (LiDAR and photogrammetry) measurement methods, active (sending/receiving a signal; GNSS, TLS, and LiDAR) and nonactive (photogrammetry) measurement methods, in situ (GNSS) and telemetry (remote measurement, TLS, LiDAR, and photogrammetry) measurement methods.

The quality of the results is closely linked to the positioning accuracy at the different stages of the technical protocols (e.g., TLS, vector, and ground control points) and therefore to GNSS, which remains an essential instrument for the calibration of the elevation data.

The graph in **Figure 6** aims at showing the complementarity of these techniques. Two parameters are taken into account: on the abscissa, the time scale/frequency of acquisition, and on the ordinate, the spatial density of the measurements, which is a key factor for the accuracy of DTMs or DSMs [47]. As the frequency of data acquisition increases and the degree of accuracy becomes more refined, finer analyses become possible. With regularly repeated high spatial resolution DSMs, it is thus

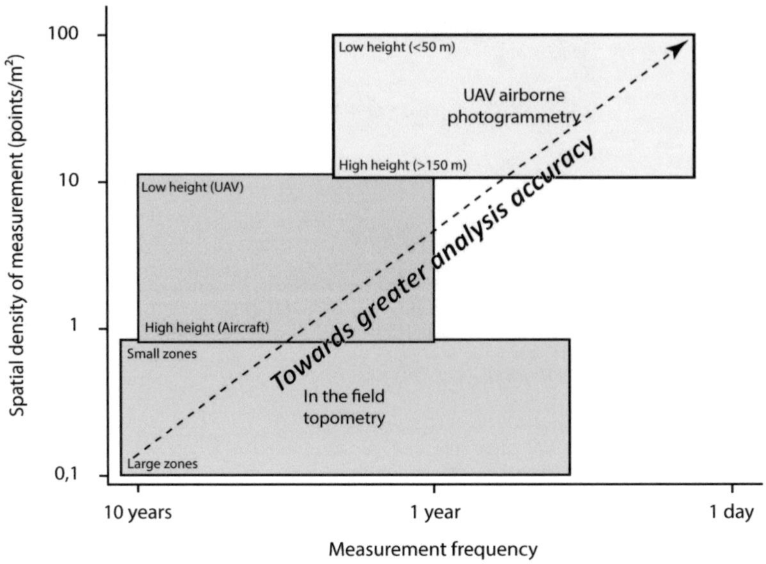


**Figure 6.**
*Spatial and temporal scales of application of the three techniques.*

possible to highlight otherwise undetectable detail changes at lower acquisition frequency and/or with coarser spatial resolution surveys (e.g., aeolian sand accumulations, see Section 2 of this chapter).

## 3. Case studies: shoreline changes on the coasts of northern France

The second part of the chapter presents examples of studies carried out on the northern coast of France in the surrounding of Dunkirk. Topographic surveys were carried out using differential GNNS, LiDAR, and airborne photogrammetry. These examples illustrate the benefits of changing from two- to three-dimensional analysis and improving spatial resolution.

### 3.1 Toward a more accurate coastline detection

The shoreline is the indicator most often used to define, map the position of the shore, and study its evolution [48]. There are several definitions of shoreline [8]. The definition may vary according to the coastal environments studied, for example, the boundary between water and sand in a microtidal environment, the base of the top of a cliff, the berm of a pebble barrier, the toe of coastal dunes, and the boundary between sand and beach top vegetation. Along the same coast, the position of these different shoreline indicators does not always coincide: for example, the limit between sand and vegetation may be located several meters away landward of the dune toe (**Figure 7**). The definition of the coastline is therefore an essential prerequisite in diachronic analyses of the shoreline.

Identifying the shoreline in two dimensions on aerial photographs is sometimes difficult. On aerial photographs, whether historical or recent, the coastline is not always clearly distinguishable. **Figure 7** shows two aerial photographs shot at two different dates (1957 and 2015), at the same scale. On the 1957 photo of poor quality, the break in slope is hardly detectable where the foredune slope is gentle. In GIS analyses, an error margin in detection must be taken into account (±*x* pixels,

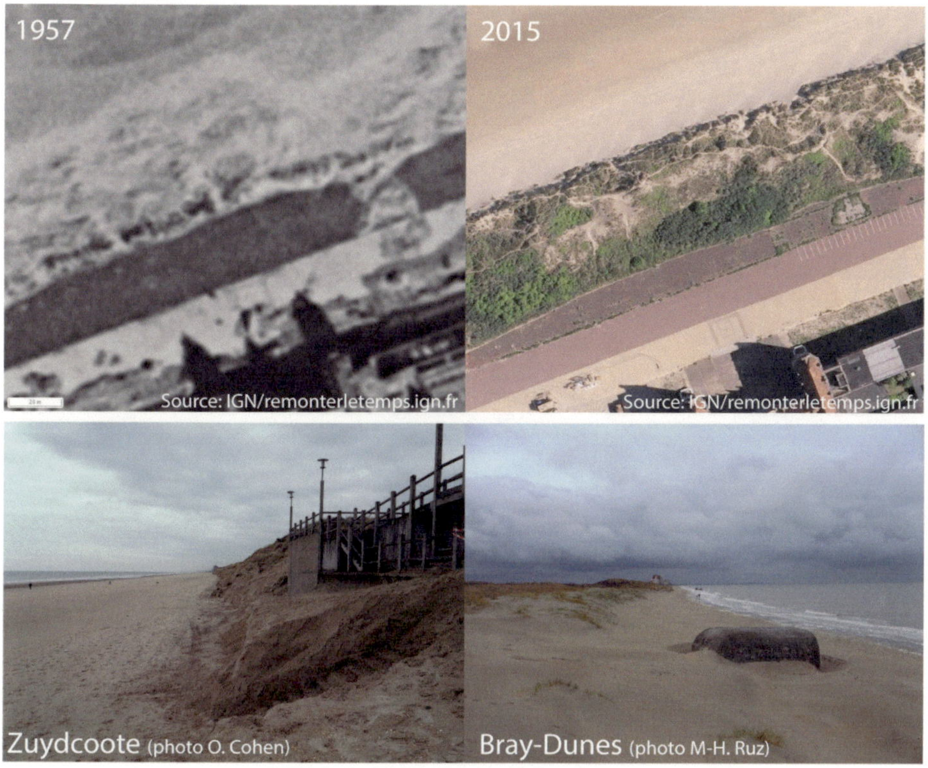


**Figure 7.**
*The complex detection of the shoreline on aerial photographs and in the field.*

thus ±*y* m, depending on the spatial resolution of the digitized photos) [49]. Stereoscopic observation of paper photos helps to identify the shoreline position, but the transfer of these observations to a computer is not convenient. Computer and photogrammetric processing of the images improves analysis (see below).

The shoreline can be measured in the field by GNSS. This requires a clear definition of what is considered as a shoreline and a good experience in field investigation. In the following examples along the coasts of northern France, the dune toe was selected. The shoreline is easily detected when the dune toe is marked by a sharp change in slope gradient (e.g., in Zuydcoote in **Figure 7**), but not when the slope between the upper beach and the dune is mild and regular (e.g., in Bray-Dunes). The measurement accuracy of GNSS is high, but in these cases, an error margin has to be attributed to the operator's interpretation of the terrain.

The shoreline can be defined by an altitude level, for example, the one reached by a distinct tide level such as the mean high spring tide. This approach is used by the French National Hydrographic and Oceanographic Service (SHOM) to define the official coastline in France, called "Histolitt," which corresponds to the highest astronomical tide (HAT). Along the coastline east of Dunkirk, for example, the HAT corresponds to an altitude of 3.787 m (French elevation datum), which has been mapped on a recent DSM obtained from an aerial photogrammetric survey with a high spatial resolution (5 cm/pixel) carried out in September 2019. The "Histolitt" shoreline has also been superimposed on the 2019 DSM showing an offset of about 11 m between the two shorelines (**Figure 8**) even if the shoreline position was defined by the same altitudinal level (3.787 m) in both cases. This difference in position between the two shorelines can be explained by the fact that the "Histolitt" shoreline was determined from a DTM with a vertical accuracy of

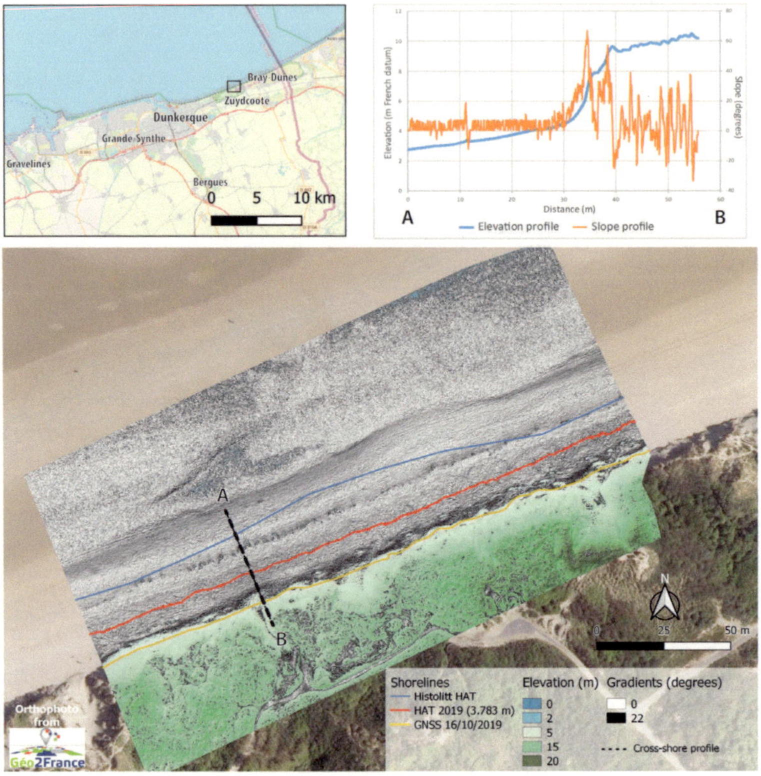


**Figure 8.**
*Example of shoreline detection on a DTM/DSM and comparison with the shoreline measured by GNSS in the field.*

50 cm calculated by photogrammetry with aerial photographs of 2004. The recent fine-scale mapping of the upper beach and coastal dunes shows that the "Histolitt" shoreline is now outdated since its position does not correspond to the present day morphology anymore. Nevertheless, even if the updated shoreline position using the location of the HAT level on the recent DSM is more consistent with the actual present day shoreline, its position still lays several meters seaward of the dune toe (**Figure 8**) that is the geomorphological expression of the upper limit of action of the highest water levels, including tides and meteorological surges related to wind and changes in atmospheric pressure [38].

The shoreline may also be mapped in a semi-automatic way by identifying in a DTM for a neat variation of the slope gradient at the dune toe [38] using the following approach:

$$\left\|\vec{g}\right\| = \sqrt{\left(\frac{\partial z}{\partial x}\right)^2 + \left(\frac{\partial z}{\partial y}\right)^2} \qquad (2)$$

where $g$ is the slope gradient, $z$ is the altitude, and $x$ and $y$ are the planar coordinates of each pixel in the DTM. **Figure 8** shows a slope gradient map calculated from a DSM computed by photogrammetry in September 2019 in Zuydcoote. The figure also exhibits two cross-sectional profiles: one is a topographic profile, and the other is a slope gradient profile in degrees. In this example, a slope gradient of 22° corresponding to the dune toe has been selected. On the map, areas with slopes between 0 and 22° range from light to dark grey. They are mainly located on the beach; there are, however, some areas with low gradients in the coastal dunes (e.g., paths). Zones with steeper slopes are not displayed on the map; therefore, the actual topography remains visible. The detection of this boundary is validated by comparison with the coastline surveyed in October 2019 with a GNNS: the differences range from 0 to 2.5 m landward, which are due to beach and dune toe erosion between September and October 2019, as a result of high water levels that had eroded the beach and the foot of the dune.

**Figure 9** illustrates a diachronic study using shorelines extracted from DTMs at different dates (2008, 2011, 2012, and 2014) on an 8 km long stretch of coast between Dunkirk and the Belgian border. The analysis of the shoreline evolution from 2008 to 2014 indicates a clear contrast between the sectors west and east of Bray-Dunes. However, the meteorological and marine conditions (winds and water levels) were the same throughout the study area [50]. The first sector is characterized by a general but moderate progradation (a few meters) from 2008 to 2011 and 2012 and from 2012 to 2014 by a mean erosion of 6 m, up to 10 m in some sectors. The second sector east of Bray-Dunes underwent accretion over all the study period with a mean seaward migration of the shoreline of 10 m: it is especially noticeable from 2008 to 2011 and 2012 and is more moderate for the last period from 2012 to 2014 where some sectors even experienced mild erosion.

### 3.2 Calculation of altimetric and volumetric changes

Switching from 2D to 3D analysis enables to detect the shoreline position more accurately. 3D analysis using DTMs or DSMs also allows volume calculation and mapping of topographic changes. On sandy coasts, this approach can be used for calculating sediment volume changes and sediment budgets and for mapping accretionary and eroding sectors that are potentially at risk [51]. Below are two examples at two different scales on the coast east of Dunkirk.

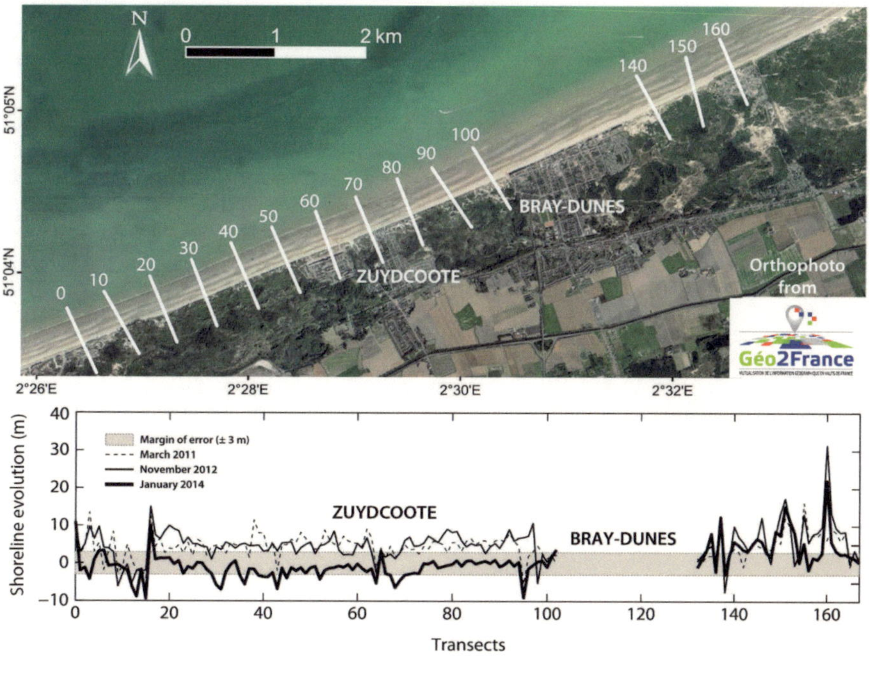


**Figure 9.**
*Evolution of the shoreline in Dunkirk and the Belgian border from 2008 to 2014 with transect location map (adapted from Ref. [38]).*

### *3.2.1 Very high spatial resolution and short-term shoreline analysis at Zuydcoote*

Four photogrammetric survey campaigns were carried out in November 2017, May 2018, September 2018, and February 2019. Very high spatial resolution DSMs were then calculated (5 cm/pixel; **Figure 10a**). The DSMs were compared in pairs (e.g., **Figure 10b**). In order to synthesize the four DSMs, a statistical analysis was also conducted. For example, a map of the annual rate of change was produced by calculating a linear regression with the pixel heights at each date.

For example, between November 2017 and May 2018, the sediment budget was negative (−5041 $m^3$, i.e., −0.15 $m^3/m^2$), 1368 $m^3$ (0.22 $m^3/m^2$) of sediment accumulated, and 6410 $m^3$ was eroded (−0.25 $m^3/m^2$). Erosion was very mainly detected on the beach and dune front. The comparison map of the DSMs (**Figure 10b**) shows a flattening of the upper intertidal bar and the upper foreshore, and the disappearance of the aeolian sand accumulation features identified on the DSM of November 2017 (**Figure 10a**). The DSM comparison also shows that the dune front has been eroding, although blowouts were filling up.

The map of the annual rate of change calculated with the four DSMs (**Figure 10c**) confirms the previous observations: erosion of the sand bar, upper beach and dune front, and accumulation in the trough landward of the upper intertidal bar and in the blowouts. The topographic and statistical cross-shore profiles (**Figure 10d**) show that the highest negative evolution rates are corresponding to the foredune front. As indicated in Section 2.2.3 of this chapter, such map must be interpreted with caution. In bare areas, the geomorphological analysis can be validated, but in the vegetated zones, the changes detected may be due to vegetation growth and not to geomorphological changes. The interpretation of these DSMs must therefore always be combined with observation of aerial photographs.

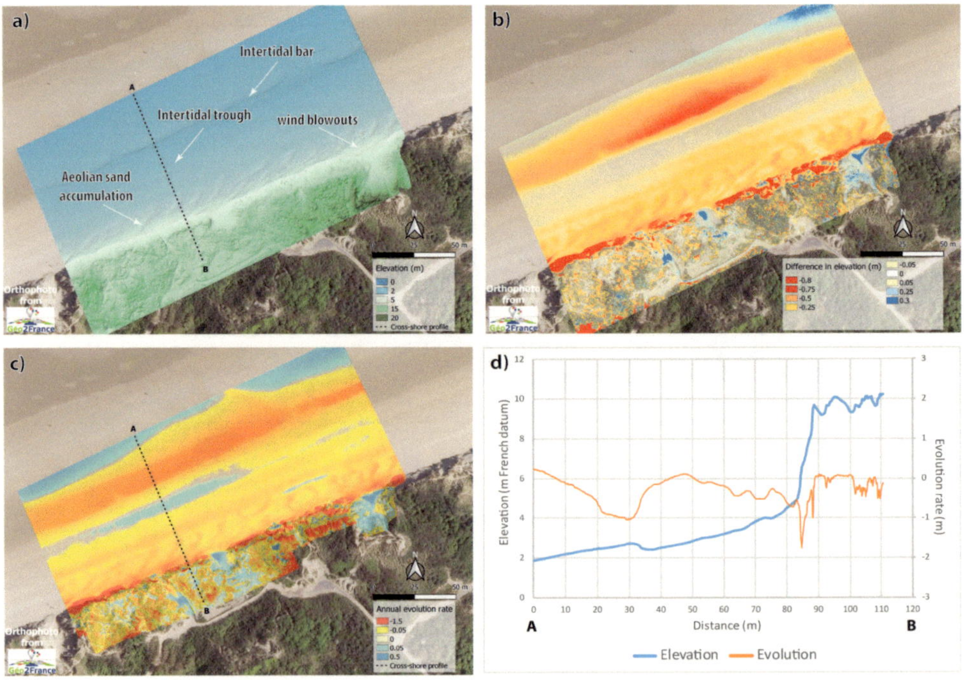


**Figure 10.**
*Examples of (a) DSM computed from photogrammetry data (November 2017), (b) evolution map (November 2017 to May 2018), (c) map of annual rate of evolution, and (d) cross-shore topographical and evolution rate profiles.*

### *3.2.2 Large-scale calculation of the sediment budget east of Dunkirk*

Four topographic survey campaigns were carried out with an airborne topographic LiDAR in May 2008, March 2011, November 2012, and January 2014 on a 8 km long coastal stretch east of Dunkirk. For each measurement campaign, a DTM with a spatial resolution of 1 m was calculated. In order to understand sediment transfer between beach and dunes, these DTMs were used to calculate sediment volume variations in the mid foreshore, upper beach, and dune (**Figure 11**). The lower foreshore limit corresponds to the minimum elevation at the time of LiDAR measurements and therefore depends on the tidal level. The upper limit of the mid foreshore is the Mean High Water level (MHW, 2.83 m French elevation datum at Dunkirk). The upper beach is the area between the MHW level and the shoreline determined using the gradient method (see Section 3.1). The seaward limit of the dunes is the shoreline, whereas the inner limit is determined by photo-interpretation according to the type of dune vegetation identified in order to exclude areas with high vegetation (e.g., sea buckthorn) where the differences between the DTM and the actual ground topography may exceed 50 cm.

**Figure 11** shows that the shoreline east of Dunkirk experienced a sediment accumulation of approximately 326 × $10^3$ $m^3$ from 2008 to 2014. Almost half of this accumulation is observed in the coastal dunes (154 × $10^3$ $m^3$), mainly in the eastern part, next to the Belgian border. The average vertical accretion is 1 m. In the whole study zone where the shoreline has been stable from 2008 to 2014, the sediment budget of the dunes is positive even if dune front erosion occurred in places. Estimates of changes in sediment volume indicate that accumulation in coastal dunes occurred primarily prior to erosive events in the fall and winter of 2013, particularly between 2008 and 2011 when a gain of more than 122 × $10^3$ $m^3$

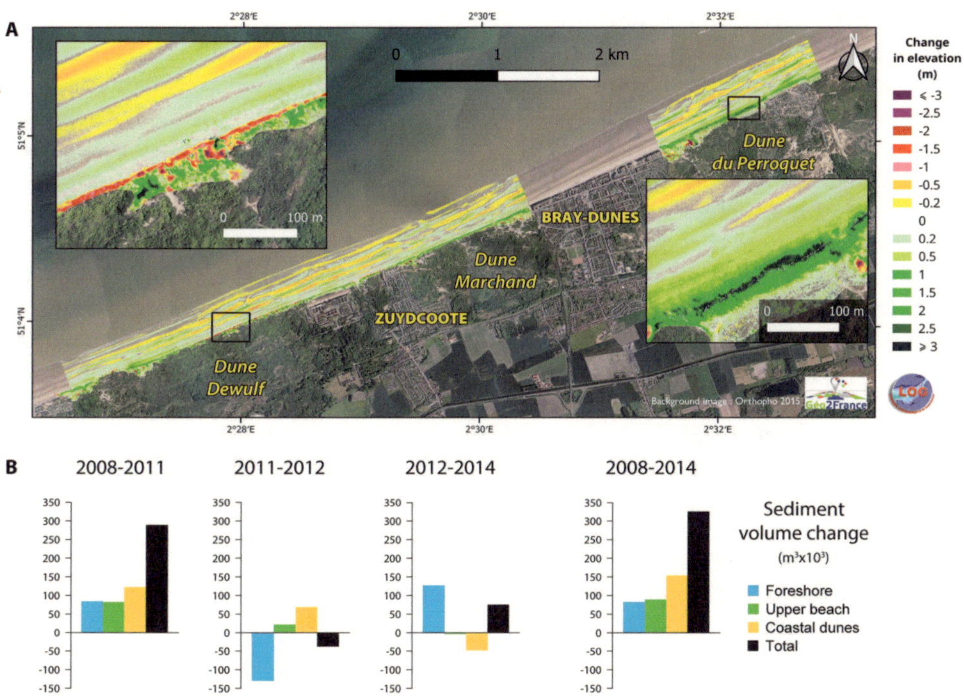


**Figure 11.**
*(A) Map of shore evolution between Dunkirk and the Belgian border from 2008 to 2014 and (B) volumetric evolution of the foreshore, upper beach, and dunes (adapted from Ref. [38]).*

of sand was observed. Sediment accumulation was about 69 × $10^3$ $m^3$ in the coastal dunes during the following period (2011–2012), which corresponds to a constant average accumulation rate of about 0.13 $m^3/m^2$/year in the dunes during both periods. Subsequently, storms in late 2013 [50] resulted in widespread coastal retreat associated with the erosion of coastal dunes west of Bray-Dunes (**Figure 11**). About 36.5 × $10^3$ $m^3$ and 15.7 × $10^3$ $m^3$ were eroded from the Dune Dewulf and the Dune Marchand, respectively, between 2012 and 2014. However, the Perroquet Dune, east of Bray-Dunes, remained fairly stable or even accumulated slightly during this storm period. Although coastal dune erosion was significant during the 2012–2014 period, a comparison of the 2008 and 2014 DTMs shows an accumulation in the dunes almost everywhere along the coast (**Figure 11A**) due to the supply of aeolian sand from the beach. Such vertical accretion is visible even where coastal dune erosion has occurred (e.g., Dune Dewulf), suggesting that wind-blown sand may also have been transported landward as the dune front eroded and retreated.

The upper part of the beach was also characterized by accretion along most of the study site, with a total gain of nearly 90 × $10^3$ $m^3$ between 2008 and 2014, with maximum accumulation (>0.5 m) measured in the eastern part of the study site (**Figure 11B**). As for the dunes, sand accumulation on the upper beach occurred mainly between 2008 and 2012, while slight erosion occurred locally between 2012 and 2014. Comparison of the 2008 and 2014 DTMs shows characteristic patterns of topographic change of intertidal bars and troughs on the foreshore. The corresponding volume changes are very limited: about +83 × $10^3$ $m^3$ over the whole foreshore, corresponding to only 0.04 $m^3/m^2$.

## 4. Conclusions

This synthesis showed that the new topographic measurement techniques implemented in coastal geomorphology over the last 25 years had enabled a clear gain in productivity in topographic measurements. The density of measurements has considerably increased, for example, for areas of several hectares, from 0.1 points/$m^2$ to 1–2 points/$m^2$ with LiDAR and several hundred points/$m^2$ with photogrammetry. This greater density of points resulted in the calculation of DTMs or DSMs with a higher spatial resolution (1 m/pixel with LiDAR measurements, 5 cm/pixel for photogrammetric measurements) on which small-scale landforms and topographic changes that could not be distinguished at a coarser resolution can be detected. With the development of these new techniques, it is therefore nowadays possible to work on shoreline morphodynamics at a very fine scale. For example, aeolian sand accumulation landforms on the upper beach and dune blowouts associated with pedestrian trampling are visible on a photogrammetric DSM but not on a LiDAR DTM (**Figure 12**: on the left, extract of the LiDAR DTM, and on the right, the photogrammetry DSM).

This improvement in spatial resolution is also associated with a decrease in measurement error margins (±20 cm horizontally and 2 cm vertically for aircraft LiDAR measurements, ±3 cm in both dimensions for low-height photogrammetric UAV surveys), which results in a higher accuracy of the DTMs produced.

**Table 1** compares the advantages and disadvantages of the three techniques. The financial costs are not discussed here.

In the coming years, coastal geomorphologists are waiting for a technique that will automatically distinguish bare areas from vegetated areas, so that DTMs can be easily calculated. The simultaneous use of LiDAR and multi- and hyperspectral sensors on board aircraft [52] or UAVs [31] is an interesting prospect. These multi- and hyperspectral images, processed with remote sensing methods, allow to map vegetation and soil moisture. This is particularly interesting for analyzing the dynamics of embryo dunes, which is closely linked to the development of pioneer vegetation [53] or monitoring the stabilization or restoration of established coastal dunes [54].

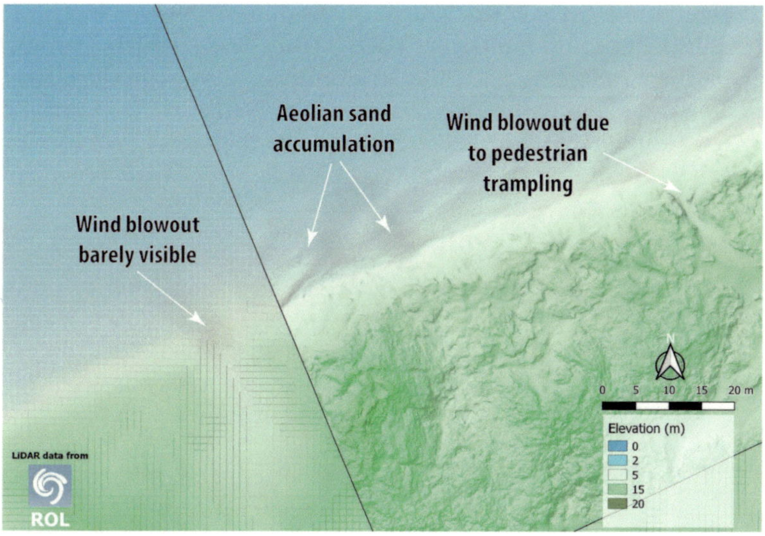


**Figure 12.**
*Comparison of DSM calculated from LiDAR data (on the left) and low-height photogrammetry (on the right).*

| Technique | Advantages | Disadvantages |
|---|---|---|
| GNSS | Measurement campaign very easy to organize<br>High repeatability<br>Not really sensitive to weather conditions<br>High accuracy measurements<br>Short processing time of the data<br>Small to medium amount of data to be stored<br>Provide DTM | Fit for small to medium study zones<br>Only provide topographical measurement<br>Low measurement density<br>Time consuming measurements |
| UAV photogrammetry | Fit for small to medium study zones (short flight time and limited range)<br>Measurement campaign easy to organize (depending on local regulations)<br>High repeatability<br>High measurement density<br>High accuracy measurement<br>Provide full orthophotographs for photointerpretation | Sensitive to weather conditions<br>Long processing time of the data and high computing power required<br>Huge amount of data to be stored<br>Provide DSM, difficult to compute DTM |
| Airborne LiDAR | Fit for large study zones (extensive flight range)<br>Medium measurement density<br>Can provide DTM | Measurement campaign difficult to organize<br>Low repeatability<br>Medium accuracy measurement<br>Sensitive to weather conditions<br>Long processing time of the data and high computing power required<br>Huge amount of data to be stored |

**Table 1.**
*Advantages and disadvantages of the three techniques.*

## Acknowledgements

The LiDAR surveys mentioned in this chapter were carried out within the framework of the interregional project CLAREC (2008–2013) coordinated by the University of Caen (France). The 2017 LiDAR data were provided by the Interregional Coastal Monitoring Network (Réseau d'Observation du Littoral Normandie Hauts-de-France ROL). The photogrammetric survey instruments were acquired, thanks to an equipment grant (BQR) from Université du Littoral Côte d'Opale. Shoreline monitoring and photogrammetric surveys on the coastline east of Dunkirk are carried out within the framework of National Observation Service DYNALIT (www.dynalit.fr). The authors would like to thank Denis Marin for his help in the realization of several figures.

## Author details

Olivier Cohen* and Arnaud Héquette
Oceanology and Geosciences Laboratory, Université du Littoral Côte d'Opale, Université de Lille, CNRS, UMR 8187, Wimereux, France

*Address all correspondence to: olivier.cohen@univ-littoral.fr

## References

[1] Masselink G, Hughes MG. An Introduction to Coastal Processes and Geomorphology. 2nd ed. London: Routledge; 2014. p. 432

[2] Davidson-Arnott R, Bauer B, Houser C. Introduction to Coastal Processes and Geomorphology. Cambridge: Cambridge University Press; 2019. p. 541

[3] Cohen O, Gardel A. De l'analyse des cartes anciennes à l'altimétrie laser aéroportée: L'évolution des techniques de suivi du trait de côte. Exemples sur la Côte d'Opale. In: Schmitt F, editor. Observation des écosystèmes marin et terrestre de la Côte d'Opale: Du naturalisme à l'écologie. Paris: Union des Océanographes Français; 2011. pp. 153-169

[4] Cohen O. La baie de Fréjus: 2000 ans d'évolution du rivage. Mappemonde. 1997;**1**(1):6-12

[5] Fenster MS, Dolan R, Elder JF. A new method for predicting shoreline positions from historical data. Journal of Coastal Research. 1993;**9**(1):147-171

[6] Hesp PA, Ruz M-H, Hequette A, Marin D, Miot da Silva G. Geomorphology and dynamics of a traveling cuspate foreland, Authie estuary, France. Geomorphology. 2016;**254**: 104-120

[7] Zemmour A. Étude de l'évolution des littoraux dunaires de la Côte d'Opale à différentes échelles de temps: Analyse de leur capacité de régénération post-tempête [PhD thesis]. Université du Littoral Côte d'Opale; 2019. Available from: https://tel.archives-ouvertes.fr/tel-02270709

[8] Boak EH, Turner IL. Shoreline definition and detection: A review. Journal of Coastal Research. 2005;**214**:688-703

[9] Perosanz F. GNSS: A revolution for precise geopositioning. Comptes Rendus Physique. 2019;**20**(3):171-175

[10] Cheng H-Q, Chen J-Y. Adapting cities to sea level rise: A perspective from Chinese deltas. Advances in Climate Change Research. 2017;**8**(2):130-136

[11] Donnellan A, Arrowsmith R, DeLong S. Spatio-temporal mapping of plate boundary faults in California using geodetic imaging. Geosciences (Switzerland). 2017;**7**(1):26

[12] Chandler BMP, Lovell H, Boston CM, Lukas S, Barr ID, Benediktsson ÍÖ, et al. Glacial geomorphological mapping: A review of approaches and frameworks for best practice. Earth-Science Reviews. 2018;**185**:806-846

[13] Telling J, Lyda A, Hartzell P, Glennie C. Review of Earth science research using terrestrial laser scanning. Earth-Science Reviews. 2017;**169**:35-68

[14] Eitel JUH, Höfle B, Vierling LA, Abellán A, Asner GP, Deems JS, et al. Beyond 3-D: The new spectrum of lidar applications for earth and ecological sciences. Remote Sensing of Environment. 2016;**186**:372-392

[15] Crown DA, Anderson SW, Finnegan DC, LeWinter AL, Ramsey MS. Topographic and thermal investigations of active pahoehoe lava flows: Implications for planetary volcanic processes from terrestrial analogue studies. In: 44th Lunar and Planetary Science Conference. Vol. 44. 2013. p. 2184

[16] Wulder MA, White JC, Nelson RF, Næsset E, Ørka HO, Coops NC, et al. Lidar sampling for large-area forest characterization: A review. Remote Sensing of Environment. 2012;**121**:196-209

[17] Beland M, Parker G, Sparrow B, Harding D, Chasmer L, Phinn S, et al. On promoting the use of lidar systems in forest ecosystem research. Forest Ecology and Management. 2019;**450**:117484

[18] Verma AK, Bourke MC. A method based on structure-from-motion photogrammetry to generate sub-millimetre-resolution digital elevation models for investigating rock breakdown features. Earth Surf Dynamics. 2019;7(1):45-66

[19] Westoby MJ, Brasington J, Glasser NF, Hambrey MJ, Reynolds JM. 'Structure-from-Motion' photogrammetry: A low-cost, effective tool for geoscience applications. Geomorphology. 2012;**179**: 300-314

[20] Stöcker C, Eltner A, Karrasch P. Measuring gullies by synergetic application of UAV and close range photogrammetry—A case study from Andalusia, Spain. CATENA. 2015;**132**:1-11

[21] Leon JX, Roelfsema CM, Saunders MI, Phinn SR. Measuring coral reef terrain roughness using 'Structure-from-Motion' close-range photogrammetry. Geomorphology. 2015;**242**:21-28

[22] Ventura D, Bonifazi A, Gravina MF, Ardizzone GD. Unmanned aerial systems (UASs) for environmental monitoring: A review with applications in coastal habitats. In: Mejia ODL, Gomez JAE, editors. Aerial Robots—Aerodynamics, Control and Applications. Rijeka: IntechOpen; 2017. Available from: http://www.intechopen.com/books/aerial-robots-aerodynamics-control-and-applications/unmanned-aerial-systems-uass-for-environmental-monitoring-a-review-with-applications-in-coastal-habi

[23] Ventura D, Bruno M, Jona Lasinio G, Belluscio A, Ardizzone G. A low-cost drone based application for identifying and mapping of coastal fish nursery grounds. Estuarine, Coastal and Shelf Science. 2016;**171**:85-98

[24] Müller J, Gärtner-Roer I, Thee P, Ginzler C. Accuracy assessment of airborne photogrammetrically derived high-resolution digital elevation models in a high mountain environment. ISPRS Journal of Photogrammetry and Remote Sensing. 2014;**98**:58-69

[25] Raparelli E, Bajocco S. A bibliometric analysis on the use of unmanned aerial vehicles in agricultural and forestry studies. International Journal of Remote Sensing. 2019;**40**(24):9070-9083

[26] Lillycrop WJ, Estep LL. Generational advancements in coastal surveying, mapping. Sea Technology. 1995;**36**(6):10-15

[27] O'Neal MA. Terrestrial laser scanner surveying in coastal settings. Coastal Research Library. 2014;**9**:65-76

[28] Mazzanti P, Schilirò L, Martino S, Antonielli B, Brizi E, Brunetti A, et al. The contribution of terrestrial laser scanning to the analysis of cliff slope stability in Sugano (Central Italy). Remote Sensing. 2018;**10**(9):3

[29] Letortu P, Costa S, Maquaire O, Davidson R. Marine and subaerial controls of coastal chalk cliff erosion in Normandy (France) based on a 7-year laser scanner monitoring. Geomorphology. 2019;**335**:76-91

[30] Lin Y-C, Cheng Y-T, Zhou T, Ravi R, Hasheminasab SM, Flatt JE, et al. Evaluation of UAV LiDAR for mapping coastal environments. Remote Sensing. 2019;**11**(24):32

[31] Suo C, McGovern E, Gilmer A. Coastal dune vegetation mapping using a multispectral sensor mounted on an UAS. Remote Sensing. 2019;**11**(15)

[32] Boeser SM. Global positioning systems (GPS). In: Finkl CW, Makowski C, editors. Encyclopedia of Coastal Science [Internet]. Cham: Springer International Publishing; 2019. pp. 905-907

[33] Brock JC, Purkis SJ. The emerging role of lidar remote sensing in coastal research and resource management. Journal of Coastal Research. 2009;**10053**:1-5

[34] Debaine F, Robin M. A new GIS modelling of coastal dune protection services against physical coastal hazards. Ocean & Coastal Management. 2012;**63**:43-54

[35] Le Mauff B, Juigner M, Ba A, Robin M, Launeau P, Fattal P. Coastal monitoring solutions of the geomorphological response of beach-dune systems using multi-temporal LiDAR datasets (Vendée coast, France). Geomorphology. 2018;**304**:121-140

[36] Shaw L, Helmholz P, Belton D, Addy N. Comparison of UAV LiDAR and imagery for beach monitoring. International Archives of the Photogrammetry, Remote Sensing and Spatial Information Sciences. 2019;**XLII-2/W13**:589-596

[37] Levoy F, Anthony EJ, Monfort O, Robin N, Bretel P. Formation and migration of transverse bars along a tidal sandy coast deduced from multi-temporal Lidar datasets. Marine Geology. 2013;**342**:39-52

[38] Crapoulet A, Héquette A, Marin D, Levoy F, Bretel P. Variations in the response of the dune coast of northern France to major storms as a function of available beach sediment volume: Response of Dune Coast to Storm as a Function of Beach Sediment Volume. Earth Surface Processes and Landforms. 2017;**42**(11):1603-1622

[39] Thieler ER, Hapke CJ. Photogrammetry. In: Finkl CW, Makowski C, editors. Encyclopedia of Coastal Science [Internet]. Cham: Springer International Publishing; 2019. pp. 1367-1374

[40] Hapke C, Richmond B. Monitoring beach morphology changes using small-format aerial photography and digital softcopy photogrammetry. Environmental Geosciences. 2000;7(1):32-37

[41] Brunier G, Fleury J, Anthony EJ, Gardel A, Dussouillez P. Close-range airborne structure-from-motion photogrammetry for high-resolution beach morphometric surveys: Examples from an embayed rotating beach. Geomorphology. 2016;**261**:76-88

[42] Casella E, Rovere A, Pedroncini A, Mucerino L, Casella M, Cusati LA, et al. Study of wave runup using numerical models and low-altitude aerial photogrammetry: A tool for coastal management. Estuarine, Coastal and Shelf Science. 2014;**149**:160-167

[43] Gonçalves JA, Henriques R. UAV photogrammetry for topographic monitoring of coastal areas. ISPRS Journal of Photogrammetry and Remote Sensing. 2015;**104**:101-111

[44] Turner IL, Harley MD, Drummond CD. UAVs for coastal surveying. Coastal Engineering. 2016;**114**:19-24

[45] Pikelj K, Ružić I, Ilić S, James MR, Kordić B. Implementing an efficient beach erosion monitoring system for coastal management in Croatia. Ocean & Coastal Management. 2018;**156**:223-238

[46] Laporte-Fauret Q, Marieu V, Castelle B, Michalet R, Bujan S, Rosebery D. Low-cost UAV for high-resolution and large-scale coastal dune change monitoring using photogrammetry. Journal of Marine Science and Engineering. 2019;**7**(3):63

[47] Grohmann CH, Sawakuchi AO. Influence of cell size on volume calculation using digital terrain models: A case of coastal dune fields. Geomorphology. 2013;**180-181**:130-136

[48] Pérez-Alberti A, Pires A, Chaminé HI. Shoreline and coastal terrain mapping. In: Finkl CW, Makowski C, editors. Encyclopedia of Coastal Science. Cham: Springer International Publishing; 2019. pp. 1558-1571

[49] Thieler ER, Danforth WW. Historical shoreline mapping (I): Improving techniques and reducing positioning errors. Journal of Coastal Research. 1994;**10**(3):549-563

[50] Héquette A, Ruz M-H, Zemmour A, Marin D, Cartier A, Sipka V. Alongshore variability in coastal dune erosion and post-storm recovery, Northern Coast of France. Journal of Coastal Research. 2019;**SI88**:25-45

[51] List J. Sediment budget. In: Finkl CW, Makowski C, editors. Encyclopedia of Coastal Science. Cham: Springer International Publishing; 2019. pp. 1505-1512

[52] Deronde B, Houthuys R, Debruyn W, Fransaer D, Lancker VV, Henriet J-P. Use of airborne hyperspectral data and laserscan data to study beach morphodynamics along the Belgian Coast. Journal of Coastal Research. 2006;**225**:1108-1117

[53] Ruz M-H, Héquette A, Marin D, Sipka V, Crapoulet A, Cartier A. Development of an incipient foredune field along a prograding macrotidal shoreline, northern France. Géomorphologie: Relief, Processus, Environnement. 2017;**23**(1):37-50

[54] Martinez LM, Gallego-Fernández JB, Hesp PA, editors. Restoration of Coastal Dunes. Berlin Heidelberg: Springer-Verlag; 2013. p. 347

Section 3

# Biological Sciences

Chapter 6

# Spatial and Temporal Variability Regarding Forest: From Tree to the Landscape

*João Carvalho, Manuela Magalhães and Selma Pena*

## Abstract

Spatial and temporal variability in forest has become a topic attracting great attention regarding the role of the forest ecosystems in biogeochemical cycles, climate change and biological diversity and in human society. Advances in the natural sciences have brought insights into and a better understanding about the patterns and processes at different spatial and temporal scales. At the same time, this supports a better management of the forest ecosystems and landscapes. Variability from the tree level to the landscape is addressed. Tree characteristics and functions, forest stand dynamics and ecological succession to forest landscape ecology are put together, considering their interrelations and dependencies. Managing forest stands and variability at different scales is described and discussed, including the scope of sustainability. An evaluation of forest and landscape characteristics in Portugal is performed with propositions considering these different elements.

**Keywords:** forest ecosystems, forest management, landscape

## 1. Introduction: variability at different spatial and temporal scales

A given forest ecosystem is part of a landscape ecology matrix that develops as a whole and in which several processes operate in variable spatial and temporal scales. Forest dynamics and spatial variability are closely linked, involving the effects of biologic processes and external factors, which occur at a wide range of spatial scales. In turn, spatial variation of environmental conditions creates variable abiotic templates where forest communities develop. In the forest ecosystem, many aspects change in time and space, whether as a result of its own process or influenced by disturbances. The ecosystem functioning, in its various expressions, emphasizes the internal dynamics of the system in a particular state [1–3].

At the landscape level, different characteristics are involved, such as the amount of *habitat,* patch size, the landscape mosaic and connectivity, which are under the scope of biological conservation, ecological restoration, forest management, landscape ecology and land management [4]. Different management levels may be more or less dependent or related. In turn, different spatial scales can also be considered (local, regional and national).

The spatial pattern of forest patches involves elements such as size, quantity, type, proportion, shape and connectivity [5]. The landscape pattern may affect the

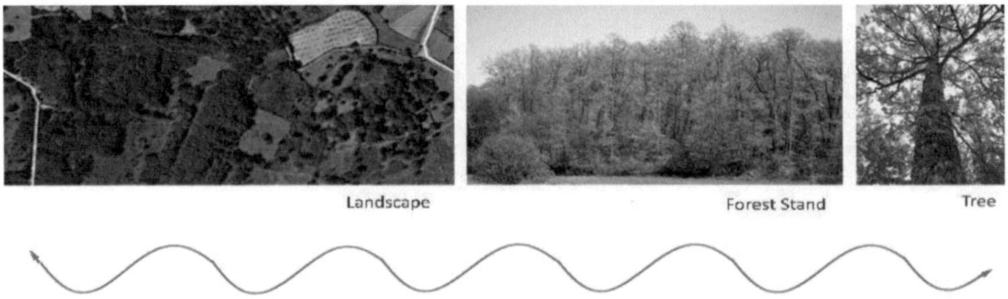


| Spatial scale | Main characteristics |
|---|---|
| Landscape ↕ | Mosaic of soil use. Spatial variation of site conditions. Different forest ecosystems. Different forest stands and interventions. Interaction of ecosystem processes. |
| Forest stand ↕ | Characteristics of the forest stand. Site conditions and management options considering ecosystem functions and level of integration. |
| Tree | Tree origin, age or size and species. Tree characteristics and functions. |

**Figure 1.**
*Forest management considering different spatial scales: tree, forest stand and landscape. Fluxes and interconnections occur among these spatial scales.*

ecological processes and the disturbances regime [6]. The spatial heterogeneity in forested landscapes may derive from different causes, including changes in ecological conditions, biotic interactions, developmental stages, land uses and disturbances.

In this chapter, the forest is assessed considering the variability at different spatial scales: the landscape, the forest population and the tree (**Figure 1**). A forest, of different size, is viewed as a part of the landscape, which might be structured according to a physiographic or natural model. A forest stand is considered a management unit that can be differentiated with a certain number of site and vegetal characteristics.

The understanding of the forest ecosystem's functioning and dynamics has improved over the past decades [7–13]. A better knowledge of their dynamics is important to forestry, providing a better decision support of the most appropriate practices to achieve certain objectives.

## 2. Disturbances and forest ecosystems dynamics

A better understanding of the forest ecosystem dynamics has allowed a broader comprehension about the influence of disturbances in the development of the forest stands and the landscape, therefore supporting the most appropriate forest management decisions.

All forest ecosystems are subject to disturbances, which may be of a different type and affect their characteristics and functioning. Natural disturbances are part of the dynamics of a forest ecosystem. The role of different disturbances, both spatial and temporal, is recognized as part of the forest development. Not only small disturbances are considered, but also major disturbances and even climate change, with their specific characteristics and occurrence.

A disturbance is any event that affects or disrupts a particular ecological process or ecosystem development; modifies the population structure; and changes the availability of a particular resource or the physical environment [6, 10, 14, 15]. A disturbance may be essentially described according to its type, frequency, magnitude or severity, extent and return period. The relative importance of each disturbance varies according to their characteristics and the type of forest.

The disturbance *type* is one of the most important characteristics of a disturbance regime. Disturbances may be biotic or abiotic, natural or anthropogenic, as well as endogenous or exogenous. Endogenous disorders are an integral part of the autogenic ecosystem development process. The potential to create endogenous disturbances varies with the species, the forest stand and development. The control and intervention on destabilizing forces are important for the development and stability, and the ability to minimize certain effects can be assessed, such as loss of water and nutrients.

The *magnitude* may affect more or less the existing plant mass. Some disturbances may destroy all vegetation, while others may leave some trees or other vegetation, which will influence the recovery process, depending on the number of remaining trees, species and position in the canopy. Major exogenous disturbances (fire, storm and clearcutting) result in a reduction or elimination of primary production and have different consequences in terms of biomass and export nutrients. A major disturbance may have an appreciable effect on the subsequent development of the forest ecosystem. For example, a fire may destroy a large part or all of the biomass and suppress primary production. At the same time, nutrients removed by volatilization and leaching can increase soil erosion. A clearcutting also removes a significant amount of nutrients present in the exploited material. A storm with the loss of many trees affects also the primary production; however, the biomass may remain in the system. Soil erosion, which can occur as a result of intensive logging, soil tillage or fire, has a strong negative impact on the ecosystem. A clearcutting or fire leads to the destruction of important hydrological, nutritional and biological soil properties. A clearcutting, particularly on steep slopes and thin soils, may lead to long-time changes on soil structure and the ecosystem biogeochemistry.

The disturbances *frequency* can be relatively variable and depend on the influence of various factors, both natural and anthropogenic. Typically, larger scale natural disturbances occur over longer periods of time. Disturbances may occur regularly or irregularly in time and space, which will be reflected in the stand characteristics and development. The *time* and *duration* of a disturbance are also important characteristics that may affect the ecosystem response.

The disturbance *extent* influences on the composition and structure of the stand, affecting the microclimatic conditions and colonization capacity from the surrounding areas. A disturbance may intervene at wide range of spatial scales (tree, stand and landscape).

Other relevant elements to consider in stand development are related to the initial conditions after a disturbance, the residual material and stand structural characteristics. A disturbance can create gaps of different size and shape, which may affect stand characteristics and dynamics. A given disturbance may affect different tropic and biological levels. The ecological effects due to disturbances and vegetation development vary with species. The resilience and the community type that are established after a disturbance are highly dependent on the ecosystem characteristics, the site conditions and the species that survive after a disturbance. Plants have different anatomical and physiological characteristics, with different adaptation and regeneration mechanisms, which allow them to face and survive certain disturbances.

A group of trees that develops after a disturbance is designated in some literature as *cohort* [9, 16]. The age range of the cohort may vary according to the extension temporal occupation. The cohort may also be referred to as singular or multiple if it results from one or more disturbance events. At the landscape level, different forestland uses may occur depending on the disturbance regime, the species, site conditions and objectives. A mosaic can be established with different compositions and structures formed by cohorts with different characteristics. Small stands do not behave like large stands since the edge is very much influenced by the adjacent area.

The direct and indirect effects of human disturbances on ecosystems and biological diversity are subject of debate and concern at various levels [7, 17–19]. Human activity has been affecting deeply forest and landscape characteristics for centuries or millennia (e.g., Ellenberg [20]). Patterns of land and forest use by man are also forms of spatial influence on ecosystems affecting various aspects such as connectivity or the edge effect on habitats. In many situations, we are witnessing a deterioration of habitats and destruction of biological balances at various territorial scales. Any effect on the ecological balance, or on any of its components, has repercussions on the entire ecological system.

## 3. The forest ecosystem in space and time

### 3.1 General forest stand spatial characteristics

There are three main forest stand characteristics that have a strong influence on spatial stand features. They are stand origin, structure and composition. Stand structure relates to the vertical stratification with different tree heights occupying

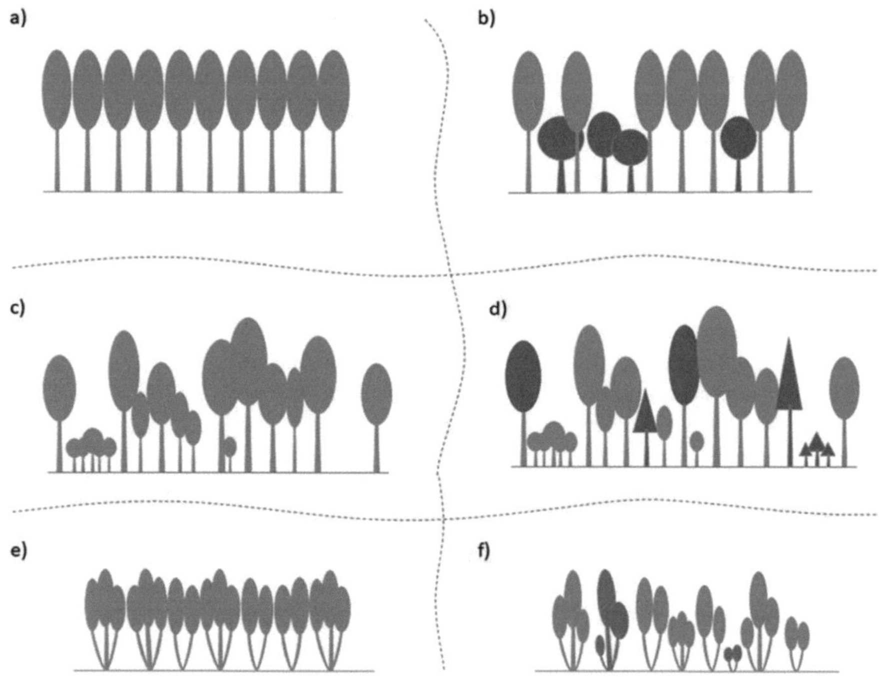


**Figure 2.**
*Forest stand origin, structure and composition have an important role on stand characteristics, influencing many functional attributes of the ecosystem. Different combinations may lead to different silvicultural systems; some are represented here: (a–d) high-forest; (e–f) coppice; (a) pure even-aged; (b) mixed two-storied; (c) pure uneven-aged; (d) mixed uneven-aged; (e) simple coppice; (f) mixed uneven-aged coppice. The figure shows the vertical and horizontal distribution (spatial pattern) of the trees within the stand.*

different canopy layers. Tree origin (seminal or vegetative) and species composition (pure or mixed) will also affect the stand stratification because of their different tree height and growth pattern. The combination and levels of stand origin, structure and composition lead to different silvicultural systems. These different forest stand components have a strong influence on the ecosystem functional processes that operate on both spatial and temporal scales (**Figure 2**). They affect stand yield, as well as forest ecologic and social functions. This also means that they have different silvicultural importance. Furthermore, their natural dispersion pattern within the stand also plays an important role and may introduce additional spatial variability (**Figure 3**).

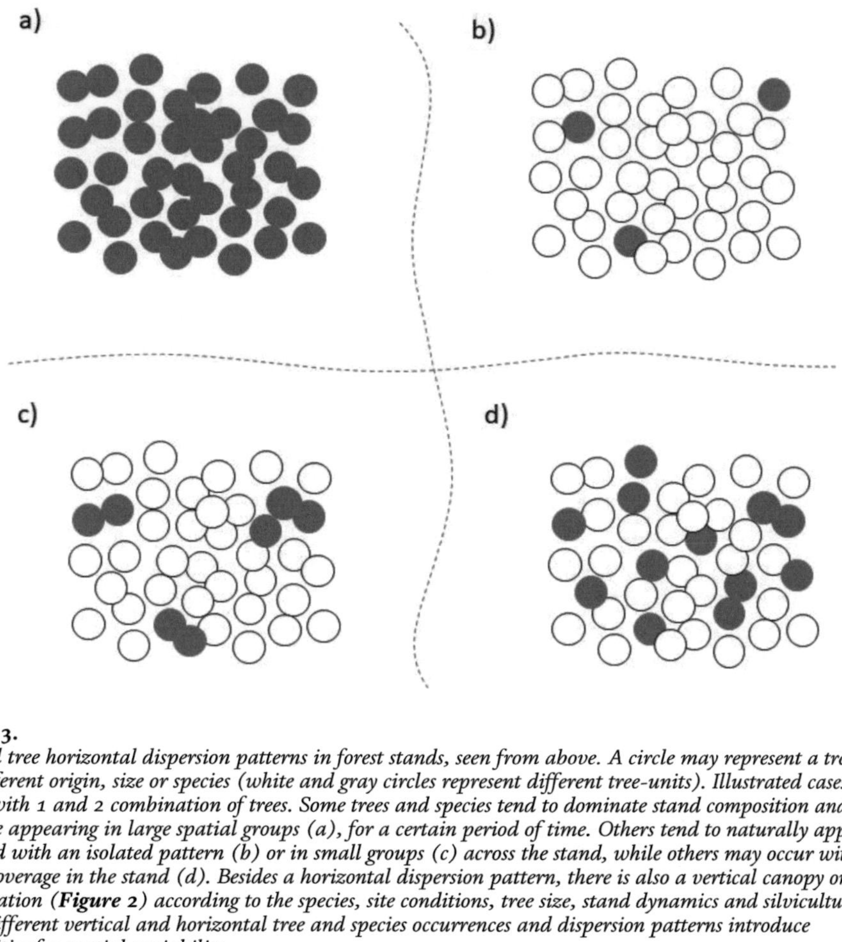


**Figure 3.**
*Natural tree horizontal dispersion patterns in forest stands, seen from above. A circle may represent a tree-unit of a different origin, size or species (white and gray circles represent different tree-units). Illustrated cases of stands with 1 and 2 combination of trees. Some trees and species tend to dominate stand composition and coverage appearing in large spatial groups (a), for a certain period of time. Others tend to naturally appear scattered with an isolated pattern (b) or in small groups (c) across the stand, while others may occur with a larger coverage in the stand (d). Besides a horizontal dispersion pattern, there is also a vertical canopy or stand stratification (**Figure 2**) according to the species, site conditions, tree size, stand dynamics and silviculture. These different vertical and horizontal tree and species occurrences and dispersion patterns introduce possibilities for spatial variability.*

### 3.2 Ecological succession: the forest ecosystem in time and space

The initial concepts concerning the ecological succession were guided to relatively predictable developmental stages of the ecosystem in general, and of plant communities in particular, in a succession of stages to a certain climax state [21]. Through ecological succession, with temporal changes in the vegetation, with biotic interaction processes, facilitation and inter- and intraspecific competition, as well as changes in habitat itself, are reached at a given time, a state of equilibrium with the climate, which results in a more stable condition and functional evolution. In this classic model of Clements, the succession consists of a predictable temporal sequence of plant communities, each modifying the environment and creating conditions for subsequent communities. This notion would be contrasted with an interpretation made by Gleason [22] where the plant communities are the result of processes of adaptation and individual development to environmental conditions, shaping the ecological continuous concept of vegetation. On the other hand, in several situations, the succession is greatly influenced by the initial plant species composition. In addition to the facilitation, other biotic processes are involved such as colonization, competition, tolerance, inhibition and survival, as well as other biotic interactions like herbivory and mutualism, which may lead to different dynamics [10, 23]. Later, other authors showed that plant communities did not behave in a simplistic way as postulated by Clements, but where the environment factors could give rise to different pathways and climax states in a given climatic region [7, 10, 17, 24–26]. On the other hand, the climax state was seen as a relatively stable equilibrium condition. Currently, the succession is mostly understood as a dynamic process of re-equilibriums and adaptations, in response to external disturbances and as a result of internal development processes of the ecosystem. The concept of succession has thus become more complex, where the prediction on the vegetation and ecosystem dynamics requires local specific information about the site characteristics, the type of disturbance, the composition and biology of the species. Many of the initial concepts included equilibrium characteristics related to the flows of energy and matter, tropic interactions and population dynamics. Complementary and alternative approaches developed concepts related to the temporal and spatial variability, the nonlinear dynamics and complex systems. On the other hand, the ecosystems are subjected to changes and adaptive processes of wider temporal scales as well as related with climate variations [11].

In addition to the temporal aspects, successional processes at the landscape scale with mosaic dynamics are also important to be considered [4, 6, 27]. In this sense, both the community and the ecosystem are landscape properties responding to changes in environmental gradients. One feature is the occurrence of successional stages across the landscape and time, gaps and patches of different sizes and trees of different growing stages within a stand.

The concept of *forest ecosystem dynamics* covers several notions, namely: the ecosystem is an open system; the ecosystems and landscapes are dynamic; the disturbance is a critical element of the system; the ecosystem is controlled by biotic and physical processes that occur at different spatial and temporal scales with levels of biological hierarchy; the succession does not necessarily follow the same pathway and ends at the same point of equilibrium; the spatial pattern is important for biological diversity; the interaction between ecosystem processes and landscape dynamics is important for biodiversity; past and recent human activities have an impact on ecosystems currently perceived as natural [3, 6, 19, 28].

The dynamics of the forest ecosystem and the temporal and spatial heterogeneity are related. The successional processes, disturbances and changes in the site factors create a complex of situations where forest communities develop (dynamic patches), which can be more wide and not necessarily in equilibrium [14, 15, 29]. Biotic interactions are also important, as are results from herbivores or pathogens and may in some cases be crucial in the development of the forest stand. The spatial pattern of the forest can itself have a strong influence on population dynamics and ecosystem processes. For example, habitat connectivity has a major effect on the abundance and persistence of certain species [30]. Therefore, besides the attributes of a certain forest, it is also important to consider the stand landscape context.

More recently emerged notions related to complex systems linked to the ecosystem dynamics. Profound changes may occur from small variations of the initial conditions. In sensitive systems, small changes to the initial conditions can result in large changes as the system evolves [10, 11]. The disturbances and heterogeneity are interdependent factors, creating opportunities for recolonization.

The different vegetation components of a forest stand are important for the ecosystem functioning. Feedback processes are also involved, which allow the development of self-regulation mechanisms. For example, less visible organisms (e.g., fungi) play important functions such as the formation of a good soil (decomposition, recycling of nutrients and formation of humic compounds), in a variety of biotic relationships indispensable for the ecosystem functioning. The temporal and spatial fluctuations, as well as the connections, are important aspects of the forest ecosystem dynamics.

The resistance and resilience concepts are related to the ecosystem dynamics, with their ability to absorb disturbances and recover to a given state. Some studies show that complexity offers greater stability to the ecosystem [10, 31–33]. The multiple interrelationships between a population and the community contribute to stability situations. The complex adaptive systems take into account the diversity and heterogeneity. They promote self-regulation, in which the reciprocal interactions within the system between the structure and processes contribute to the regulation, organization and dynamics. Different initial conditions are directed for a stable situation, becoming relatively robust for certain disturbances, where the system components adapt. On the other hand, in simplified or unstable systems, small disturbances may have a destabilizing and destructive effect.

### 3.3 Forest stand development stages

The forest development stages provide an idea about the changes that operate on a forest stand as regards the structure, composition and ecosystem processes associated with the dynamics of a population of trees. These stages seek to provide a general framework in which certain conditions and procedures are more prevalent. They occur successively and may also involve processes that operate at different sizes and moments in the stand. These variations are related to the concept of dynamic equilibrium of the forest ecosystem. The ecosystem functionality will be linked to structural, compositional and population dynamics characteristics.

Several authors (West et al. [7]; Oliver and Larson [9]; Spies [34]) recognized the following stages in the development process of a forest stand:

- Establishment, initiation or re-organization stage
- Stem exclusion stage
- Transition or understory re-initiation stage
- Old-growth or shifting mosaic stage

**Figure 4** shows the evolution of total biomass at different development stages, after a clearcutting. In the re-organization stage, a loss of total biomass occurs, where growth and living biomass accumulation begin. In the stem exclusion stage, the ecosystem accumulates biomass to a certain point. In the transition stage, the total biomass decreases slightly until it stops in a fluctuating way in the old-growth or durable mosaic stage. The biomass reaches a maximum at the beginning of the transition stage, decreasing and stabilizing subsequently as a result of mortality of dominant trees that are replaced by smaller trees. Carbon retention in the living and dead components of the ecosystem may also reach a maximum at this stage. Throughout these stages, a development of the stand structure occurs through

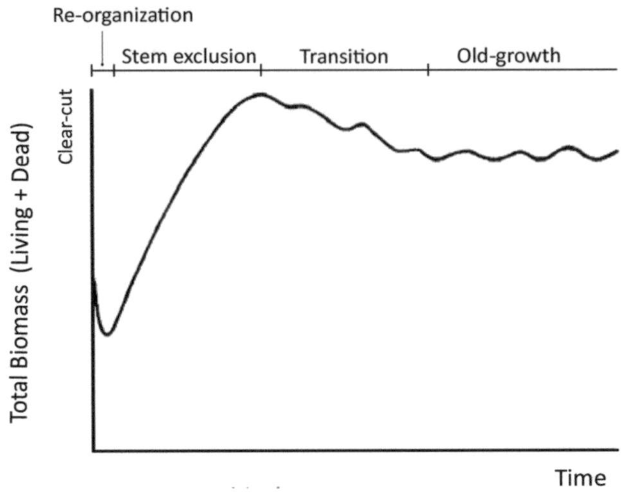


**Figure 4.**
*Evolution of total biomass at different stages of stand development, after a clearcutting (adapted from Bormann and Likens [35]). Stages are delimited by changes in total biomass (living and dead biomass), assuming a natural development without exogenous disturbances.*

different ways depending on the species, site conditions and the dynamics of the stand itself.

Old-growth stands receive a special attention from the point of view of ecology, conservation and forestry, addressing aspects related to biocenosis, genetics, ecosystem, management and the landscape. The various definitions on this stage or stand type show the diversity of interest. In its definition, structural elements as well as the state of the development process should be considered. According to Spies [34], it is a forest ecosystem distinguished by the presence of very old trees for the particular soil and climate conditions in which it occurs, showing certain morphological and growth characteristics.

The first phase occurs after the occurrence of a disturbance in which new individuals are established. The structural complexity varies depending on the type of disturbance and the present biological elements. Relative fast changes occur on the forest environment, the competition level, the species dominance and the population structure. At this stage, there is a great diversity of species, which may decrease as the space is being colonized by trees [36]. A severe disturbance leads to a regression of the forest ecosystem to an earlier stage of ecological succession. The development pattern after this event is also greatly influenced by the present floristic composition. The relative importance of species can vary in time and space according to the reproduction and growth strategies, and modifications of species dominance may happen.

At this stage, as a result of the disturbance, there is a loss of biotic regulation of the system. Hydrological and biogeochemical parameters are changed and deregulated. In turn, there is a temporary increase in the availability of resources, as well as an increase in solar radiation at the soil surface. The clearcutting has a strong effect on many ecosystem processes and greatly modifies the regulation ability of radiation energy flow as well as the hydrological and nutrient cycles. With the removal of the forest cover, the microenvironment is affected, with an increase of the soil temperature as well as various processes such as the absorption of nutrients and water, transpiration, the absorption and reflection of solar radiation, the primary production and the production of litter. The system's ability to store water and nutrients is greatly affected. The decomposition of organic matter is accelerated. There is a loss of soil organic matter and an increase of soil acidity.

The restoration of vegetation leads to a progressive reduction of soil erosion. Nevertheless, some studies show that even several years after clearcutting, having attained a full canopy coverage, which may extend for 15 years or more, may still register important decrease in the soil organic matter and nutrients until full control is restored [35, 37, 38] which in turn affect attributes such as soil water holding capacity and carbon storage (**Figure 5**). A reduction in soil thickness may also occur, which may extend for a long time after a clearcutting as a result of the effects in many ecosystem processes [39, 40].

The reduction of transpiration and soil water holding capacity has a pronounced effect on the hydrological cycle. There are frequent situations where rainfall produces flooding. Situations involving the transport of particles and soil erosion become more problematic. This stage ends when the living and dead biomass are accumulated in the ecosystem and end the decline of organic matter, and the restoration of biotic regulation of the hydrology and nutrient exportation occurs.

In the stem exclusion stage, a progressive reduction of tree density typically occurs as a result of the intense competition among the trees. The more evident characteristic is a rapid accumulation of biomass with a competitive exclusion of many individuals where mortality is very much dependent on the population density. The net primary productivity can be very high and some characteristics of the population, such as leaf area, can reach a maximum. The loss of nutrients at this stage is lower due to intensive use of existing resources. The stand instability against unfavorable atmospheric events may be higher due to a high population density. The tree mortality is more intense particularly in the lower and intermediate classes of light-demanding species, so there may be a reduction of the species diversity compared to the first stage. At this stage, the canopy is relatively uniform and there are few gaps. Canopy openings as a result of death of individual trees are of small size. Variations in the growing space, the species, tolerance, age, genetics, competition, site characteristics and external factors influence the growth pattern. The density of the dominant trees decreases as its size increases. Compared to the previous stage, there is a better regulation capacity of the energy flow, hydrological and nutrients through the biotic and abiotic components of the ecosystem. Another important feature is the regulation of the chemical composition of the drain water.

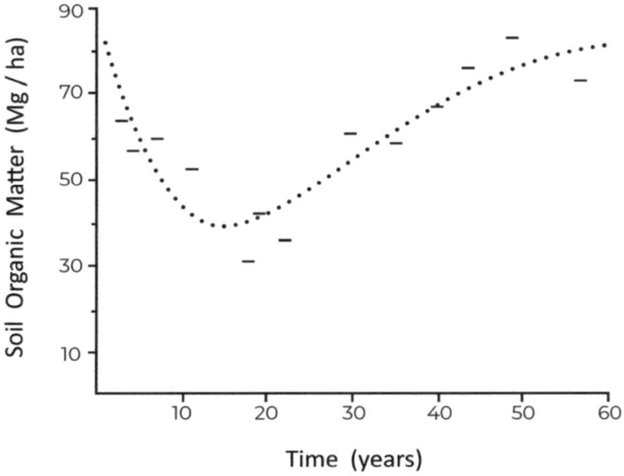


**Figure 5.**
*Following a clearcutting, on a broadleaved stand (*Acer, Betula, Fagus, Fraxinus *and* Prunus*), a degradation of the soil organic matter occurs that extends up to 15 years and a loss of 51% of the initial content. The recovery to initial values may take nearly 40 years (adapted from Covington [39]).*

In the transition stage, some authors recognize that in some types of forest there may be steps or sub-stages (understory release, maturation, early transition, and old-growth and late-transition old-growth) [41]. Two major steps may occur: a transition phase and a steady-state or shifting gap phase. Gradual changes occur in the population, the structure and development process, which together may have a very variable duration. The living biomass and diversity of forms reach a maximum. The initial group of trees disintegrates gradually, the mortality of lower trees increases, and a new group of trees may gradually grow in gaps. Some authors such as Oliver and Larson [9] recognize a stage of re-initiation, where a new group of trees grows in the understory. A transitional phase to an old-growth is developed where initial trees are also present. Compared to the previous stage, a progressive decrease of total biomass up to a more or less stable level occurs. The amount of dead wood tends to be more or less stable, fluctuating around a certain value. Species diversity increases where endogenous disturbances become more important. The death of trees leads to changes in microclimate conditions and resources. Canopy gaps promote the availability of resources, which are used by pre-existing or new regeneration. The occupation that occurs will promote stand stratification. This stage presents a great stability and resilience of the ecosystem to destabilizing events.

The old-growth or shifting mosaic stage is characterized by a pattern of relatively small disturbances, resulting in gaps of different sizes, which create conditions for the establishment of new trees and growth of trees from the lower and middle layers. The aggregation and dynamics of these small disturbances, and tree response from a larger spatial scale, result in a very small change state. Hence, some authors also designate this stage as a durable state or dynamic mosaic [7, 35]. A longer period of time is required for the establishment and development of this stage. In most cases, it is not present or occurs incompletely as a result of logging activities or frequent disturbances. The disturbance pattern, climate fluctuations and other external factors affect also the stand development. Some structural features are present in this stage, such as old and large trees, dead standing and down trees, trees of variable size and age, and a diverse understory. According to several authors, the total biomass remains relatively stable with little fluctuations over time. Slight variations of biomass occur between different parts of the ecosystem, the living biomass, dead wood, floor organic matter and the soil organic matter, with development interactions and balances. The environment conditions do not differ much from the last transition stage.

At this stage, there is a progressive elimination of old dominant trees and the development of dominant trees of different ages. These processes may lead to the formation of a population with a high degree of differentiation and structure. The stand may contain different tree species, which develop in different microclimate conditions. The stand may present a considerable biological diversity. At this stage, there is also a horizontal diversification, with different structural units.

The diversity of habitats increases as the ecosystem includes various states of development. Certain species have a greater abundance and development at this stage, due to their low rate of colonization and growth, as with certain lichens, fungi and tree species. Many species are dependent for their survival of dead wood or other structural features of the stand only present in this development stage.

Regarding the hydrological and biogeochemical cycles, dynamic oscillations occur as a result of occasional disturbances. Nevertheless, the ecosystem taken as a whole is relatively stable and resilient through different processes. There is a stabilization of the total biomass and storage capacity, regulating the export of nutrients. This stage corresponds to a relative equilibrium condition in relation to growth and mortality, the hydrological and biogeochemical state. The forest ecosystem has a great resilience, able to absorb disturbances and persist within certain limits.

## 4. Tree-level variability

Trees have various attributes such as the species, age, size, anatomical features and the dispersion or occurrence pattern (**Figure 3**). Another dimension is related to the function that a given tree may have depending on its characteristics, location and silvicultural options (**Figure 6**). This aspect introduces an additional element of variability. These functions may be related to aspects such as: production; protection; education; regeneration; biodiversity; and aesthetics. In turn, different species present distinct natural dispersion patterns. Certain species occur on an aggregate pattern, while others are more scattered.

The presence of certain trees with particular biodiversity objectives and providing tree-related microhabitat structures is also an important aspect to consider (*habitat trees*). These are living or dead trees with singular anatomical characteristics or providing ecological niches of interest to a wide range of various life forms including rare and endangered species. Anatomical features such as tree size, snags, branching variations, broken top, dead branches, stem cracks, fork crack, rotten wood and stem cavities are of interest. In some cases, these might be remarkable and monumental or veteran trees. Different studies have shown that the presence of large trees, cavernous and dead trees, standing or down, has an important contribution to biodiversity [13, 42]. These microhabitats support a complex biological network, providing food, shelter and reproduction space, contributing to the ecosystem functioning. Certain species are particularly associated with these habitats, being important conservation components (e.g., saproxylic fungi and insects).

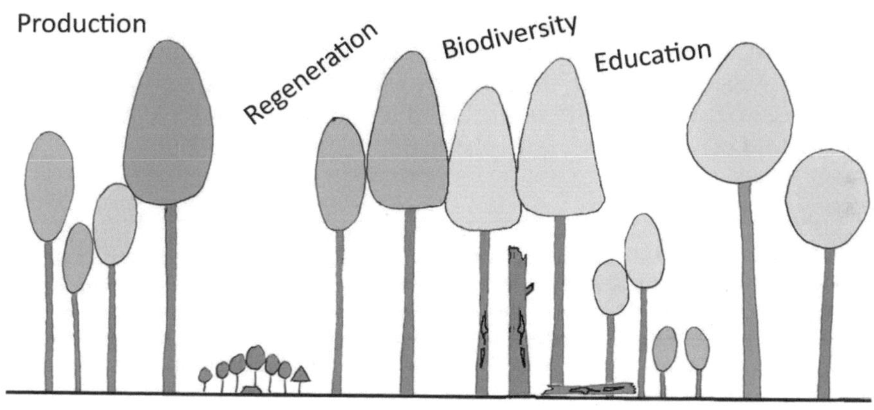


**Figure 6.**
*Representation of some functions attributed to trees. Diverse tree characteristics may provide different functions and variability.*

## 5. Forest ecosystems in the landscape

The forest has vital importance at the landscape scale, going far beyond the production of materials and energy, called *tradable goods*. The other functions of the forest in the landscape are the conservation of water, soil and biodiversity. These supporting ecosystem services are interrelated with other forest-related ecosystem services, such as climate regulation, bioclimatic comfort and other cultural services (landscape contemplation, recreation and cultural heritage). The quality of the ecosystem services provided depends on the principles of landscape planning, followed by the technicians and policy-makers, which will also determine the sustainability of the forest.

The adequate provision of forest ecosystem services, without disturbing the landscape equilibrium, depends on the understanding of the ecological and cultural landscape context. For this, the forest planning needs to be defined in articulation with other uses considering the river basin context. The river basin is a fundamental landscape unit of planning because everything flows in it: water, sediments, nutrients, air, through the local breezes (mountain and valley), and even man and goods. This flow of energy and materials depends on the land morphology of the river basin [43], but also ecological components, invisible or unnoticeable to an ordinary observer, such as the lithology, the characteristics of soil and the land cover types existing in the basin with different behaviors in the rainwater infiltration. This last aspect of the land cover is also crucial in the thermal and water balance of the atmosphere, because if there is a change on the land cover and land use, there is a changing of the planetary albedo, meaning a change on the reflection coefficient for solar radiation. Albedo is a crucial climate factor. Thus, climate change should be discussed in an integrated way [44] concerning the impact of land use and land cover changes.

The location of the forest and the type of species used should, therefore, be planned to take into account all these aspects, through a landscape design that articulates them in patterns of occupation (mosaics) capable of also ensuring other functions previously described, such as continuity and stand fragmentation.

The methodology that has been developed considering the integration of different components of the landscape system is being studied in the SCAPEFIRE project (**Box 1**). The conceptual approach is to include layers by objectives and then develop their spatial integration. In order to ensure the ecological sustainability of the landscape, the layers considered are (i) water conservation, (ii) soil conservation and (iii) biodiversity conservation. To these is added one more layer concerning the sustainability of the forest by itself: (iv) the prevention of rural fires. The areas from (i), (ii) and (iii) are included in the landscape ecological network [43, 45, 46] and need to be carefully planned.

The overarching goal of the SCAPEFIRE Project is to propose a landscape planning model that contributes to the prevention of rural fires, considering the ecological, economic and social sustainability of the landscape. Some Portuguese landscapes are highly combustible due to the last four decades of inadequate policies. Despite the importance of spatial planning as a core component in the rural fire prevention, mentioned in the media and the political discourse, its definition and implementation are still to be accomplished. The proposed project is based on the assumption that a paradigm shift in the land use is needed in favor of a lesser "fire-prone" and a more sustainable model. Acknowledging the current economic importance of the most fire-prone species, the aim is to create a landscape protection structure against rural fires that ensures soil, water and biodiversity conservation and socio-economic viability. This structure will be adapted to each type of landscape. In addition to the proposal for a new land-use planning model, the economic evaluation of multifunctional agroforestry systems will be carried out. Moreover, by improving and valuating native broadleaved species by their multiple goods and services, they provide with higher comparable profitability. Therefore, this project aims to integrate the sectoral themes in a landscape/land-use plan. Its main innovation lies in the transdisciplinarity that has not been usual, either in the field of rural fire research or in public and political discussion. The core project team consists of a permanent group of researchers based at LEAF/ISA/ULisbon, where coordination is located. This team will bring together researchers across several Portuguese research centers and other national institutions and the Pau Costa Foundation with extensive knowledge on fire ecology and operational fire management at landscape level. Stakeholder participation will be present throughout the Project, through a group of researchers, public entities linked to land-use planning, at national level (Directorate General of the Territory), and local (municipalities), but also owners and the Portuguese Federation of Local Development Associations (MINHA TERRA).

**Box 1.**
*Species movement.*

### 5.1 Water conservation

The main objective regarding water conservation in a river basin is to maximize its concentration time, meaning to increase its retention as long as possible before arriving into the sea through the rivers. This objective is most important in the Mediterranean climate, as in the case of Portugal, where precipitation occurs in the cold season when plants are at vegetative rest, so there is an imperative need to store rainwater in winter so that it can be used in summer. The best storage is underground, that is, in aquifers, because this prevents evaporation losses and provides better water quality due to the effects of filter and buffering capacity of the soil. In order for water to get to aquifers, it must be retained so that it has time to infiltrate.

Infiltration can be achieved in two ways: a natural mode and a forced mode, that is, with active measures in that direction. Natural infiltration requires knowledge of the combined permeability of lithology, soil and slope degree [47].

At the river basin scale, it is also essential to address the areas where infiltration is ensured, even at low permeability. It is vital to infiltrate and retain water in the headwater systems [48] and, as much as possible, in the upper third of the basin. In this regard, Molchanov [49] indicates a minimum size of 40% of the basin's afforestation area, ensuring a convenient full water flow.

Another measure to achieve water retention is the selection of vegetation species that can contribute to good soil that has water retention capacity and also to produce a highly absorbent organic layer of soil (leaf litter and humus). In this regard, Molchanov recommends a combination of hardwoods and *Cupressaceae*. The species to be used must be autochthonous, meaning that, in each case, research is needed on the best leaf litter to obtain. As for soil capable of better retaining water, it will be developing in the following point, which is a common feature of all landscape-system layers.

Other areas of the river basin where water conservation is required are the streams, their banks, and floodplains, and also the springs. The banks and springs should be lined with vegetation from the riparian gallery, from various strata, from aquatic herbaceous plants to the tree layer. Floodplains, depending on the time of year, are wetlands, or even subject to flooding. They should be reserved for suitable crops or riverside trees and never have buildings (other than small irrigation or other support infrastructures). Depending on their situation in the river basin (upstream or downstream), these areas usually do not infiltrate water, especially in the rainy season, when the lower section has already depleted the infiltration capacity. Floods that occur downstream of the basin, depending on conditions, can to a large extent be controlled or mitigated by basin planning, especially in the upper third, either with appropriate coverings or with forced measures.

### 5.2 Soil conservation

Soil plays a crucial role in the capacity to retain water. This capacity depends on the texture of the soil (coarser textures seep more water, while the finer textures retain it until the soil reaches field capacity, after which it begins to shed water and needs to be drained to support the plants) and depends on the soil structure. Only a well-structured soil can retain water through the clay-humic complexes and for that it has to have organic matter, including lignin, and a balanced microbial life, in which all microorganisms play an essential role, including fungi due to mycelia and their role in improving the conditions of nutrient use by plants.

These characteristics are usually ignored by people, but also by technicians who advocate soil-destroying cultural practices. These include soil loss by building or compaction, but also by soil tillage or tillage techniques, contributing to water evaporation and erosion, especially as the slope increases. Also, the practice of prescribed fire and the use of biocides are severely detrimental to soil quality due to the destruction they cause of their biome. The issue of erosion should be addressed in the presence of soil erosion maps [50] in order to propose the best land cover to provide pedogenesis and water infiltration. In the absence of these maps, it is well known that, among the factors involved in erosion, the slope is decisive, so depending on the soil types and the evidence of erosion, it is necessary to evaluate which slopes from which erosion control techniques should be programmed. However, on slopes greater than 25%, those soil erosion control techniques should always be considered.

### 5.3 Biodiversity conservation

Concerning the conservation of biodiversity, much has already been mentioned. With regard to the landscape, aspects that still need to be considered, apart from those for water and soil conservation, are the fragmentation and continuity necessary for the conservation of life flows (plants and animals) (**Box 2**). Continuity should be ensured in the main structural lines of the landscape (ridges and waterlines), creating links with existing forest areas. Where there is no forest, both in rural and urban areas, continuity should be achieved by partitioning the landscape through linear biodiverse structures (hedges) consisting of shrubs and tree species, depending on the functions to be obtained, in addition to biodiversity (wind protection, reduced evaporation, shading, field or path delimitation, etc.).

An important aspect about the variability and biodiversity is related to the landscape and fragmentation of the forest habitat. Many studies, including in the Mediterranean region, have shown that an excessive forest fragmentation is another element of fragility and vulnerability of the forest with adverse effects on biodiversity, economic and landscape values [51–54].

As the forest is gradually fragmented, with patches of reduced size and increasing distance, the habitat became increasingly more isolated. This has a major impact on habitat loss, on the different biotic communities, the population dynamics and processes of the forest ecosystem. Habitat connectivity has an important effect on the persistence and abundance of different species [55, 56]. The gradual fragmentation may also lead to the extinction of species of different biological groups that are more sensitive to this process. The colonization of a species results from the combination of dispersion and recruitment. Certain species of slow dispersion are affected by excessive fragmentation. For certain species, with a narrow ecological niche or limited dispersal ability, habitat reduction leads to risk of extinction of local populations. On the other hand, small fragments are more susceptible to degradation factors. In smaller fragments, the edge effect is larger.

Habitat destruction leads to biodiversity loss not only in the affected areas but also in the fragments due to the population size reduction, the disruption in the movement and interactions [52]. The functional connectivity is a crucial factor in the viability of certain populations, the dynamics and interspecific interactions (e.g., Tilman and Kareiva [57]). Species movement and dispersal, genetic exchange and other ecological flows in a given area are important for the survival and viability of many species [19]. Some studies show that as the proportion of a given habitat reduces, the colonization possibilities of the remaining fragments decrease (e.g., With et al. [42]). Fragmentation has also effects on the stand genetic variability [58]. Recovery after a disturbance will be heavily influenced by the availability of seedlings and the connectivity to existing nearby populations.

**Box 2.**
*Fragmentation of Forest Habitat.*

### 5.4 Conservation of the forest itself: prevention against rural fires

The occurrence of mega-fires in recent decades, not only in Portugal (since the 1980s), but in other countries, introduces another problem to solve in landscape planning: rural fires. Admittedly, landscape management cannot solve all occurrences, especially when they are of criminal or negligent origin. Nevertheless, it can reduce the size of the fire, curb its progression and even promote its self-extinguishment.

One of the critical components of fire behavior is known to be basin morphology, including slope, aspect and altitude [59–61]. The slopes exposed to the north, with slopes >25%, are the least burning [62]. The speed of fire progression doubles for each 10° increase in slope [63] and is reduced when it reaches the top, due to the local wind from the opposite slope. When it reaches the ridge, if it does not progress in the opposite slope, the fire begins to plow toward the lower slope, more slowly than when the slope rises. Given this pattern of fire behavior, it is essential to create a landscape fire-prevention network directly related to the watershed morphology that contains or extinguishes the fire. Agee et al. [64] propose the installation of shaded fuel breaks as low-fuel vegetation strips or areas (note that they do not correspond to the fuel management strips provided for in Portuguese law, with no vegetation and bare soil). These authors propose that these shaded fuel breaks be networked, according to the site, and say they are more efficient if they are wide and have surface fuel control bands.

The key areas of the river basin in which to intervene for this purpose are the structuring lines of the landscape-the streams and the ridges. According to Povak et al. [65], the waterlines and associated valley bottoms are more important for this purpose than the ridges. If the slope is too long, one or more fire retardant strips should be introduced downhill along the slope to avoid top-down and down-up fire [61]. To complete the structure, it is also necessary to create strips transverse to the slope. In the hillslope, the streams and the secondary ridges alternate, so it is in these secondary lines that these fire-retardant strips should be created [66].

Concerning the species to be used, there is a considerable debate about the higher or lower combustibility of species. In Portugal, *Eucalyptus globulus* Labill. and *Pinus pinaster* Aiton. have occupied the country and are currently the two species with the present main commercial value, since the industries related to the transformation of autochthonous species have practically disappeared, which discourages the owners for their use. The simple empirical observation of fires and their consequences, as well as the analyses carried out on the species that burned the most, allows to say that these two tree species are more combustible than the autochthonous tree species. From the available literature, Silva et al. [67] verified a tendency toward fire, in decreasing order of: pinewood, eucalyptus forests, broadleaf forests, unspecified coniferous forests, cork oak forests, chestnut orchards and holm oak. They also concluded that stand composition is the most important variable to explain the probability of fire. Calviño-Cancela et al. [68] also state that autochthonous species are more fire-resistant, as well as the studies concerning leaf litter combustibility [69]. In this context, it has to be admitted that species are not equally combustible and that, as might be expected, hardwoods other than eucalyptus are more fire-resistant and therefore can be regarded as fire-retardant. The landscape fire-prevention network should, therefore, be planned with different tree species, always avoiding monocultures. In this network, it is also possible to have the agricultural fields, pastures and, ultimately, voids (without shrub or tree vegetation) that, however, should be covered with herbaceous plants so as not to leave the soil uncovered and prevent its erosion.

## 6. Evaluation of the forest condition in Portugal: analysis and results

Different characteristics and variables concerning the forest of Portugal were evaluated to provide an overview of their status and condition. Data were collected considering different sources and analyzed, taking into consideration the main features related to the sustainable forest management goals and including the forest landscape features in relation to the habitat mosaic and connectivity.

An analysis considering the patch size by forest type and the land morphology was performed.

**Table 1** presents several characteristics of the forest in Portugal and its comparison with Europe.

| Variables and indicators | EU-27 | Trend | Portugal | Trend | Evaluation |
|---|---|---|---|---|---|
| Forest area – 2010 *(x 1000 ha)* | 177,757 | +2% | 3164 | −2.5% | 🟡 |
| *% land area (average)* | 38.0 | 1990–10 | 35.5 | 1995–15 | |
| *(range)* | (10.8–76.6) | | | | |
| Growing stock *(M m$^3$)* | 805.6 | +12% | 154 | −27% | 🟡 |
| *(average and range)* | (13–3466) | 1990–10 | | 1990–10 | |
| Coppice forest (*% forest area*) | 9.8 | | 27.4 | | 🟡/🟢 |
| *(average and range)* | (0–48) | | | | |
| Irregular stands *(% f. area)* | 25.8 | | 3.0 | | 🟡 |
| *(average and range)* | (0–95) | | | | |
| Tree species composition | | | | | |
| (*% forest area*) | | | | | |
| 1 species | 29 | | 50 | | 🟡 |
| 2–3 species | 51 | | 44 | | |
| ≥4 species | 20 | | 6 | | |
| Introduced species *(% f. area)* | 5.0 | +0.7% | 28.0 | +14% | 🟡 |
| *(average and range)* | (0–70) | 2000–10 | | 1995–2015 | |
| Regeneration *(% f. area)* | | | | | |
| Natural regeneration | 56 | | 53 | | 🟡/🟢 |
| Planting | 34 | | 25 | | |
| Coppice | 10 | | 22 | | |
| Carbon stock – above gr *(M t C)* | 292 | +26.8% | 102 | | 🟡 |
| *(average/country and range)* | (23–1405) | 1990–10 | | | |
| *(average t C/ha)* | 44 | | 32 | | |
| Forest functions *(% f. area)* | | | | | |
| (*primary function*) | | | | | |
| Production | 61.9 | | 64.7 | | 🟡/🟢 |
| Protection soil and water | 9.7 | | 7.5 | | |
| Conservation biodiversity | 12.2 | | 5.5 | | |
| Social services | 2.0 | | <1 | | |
| Protection area | 14.2 | | 22.3 | | |
| Fellings *(M m$^3$) (average)* | 16.9 | +19% | 9.6 | | 🟡/🟢 |
| *(average/forest area, m$^3$ ha$^{-1}$)* | 2.6 | 1990–10 | 3.0 | | |
| Products (*sum & average*) | | −9% | | | 🟡/🟢 |
| Roundwood *(M m$^3$)* | 405 (15.0) | 1990–10 | 9.6 | | |
| Mushrooms *(M kg)* | 429 (15.9) | | n/a | | |
| Fruits *(M kg)* | 376 (13.9) | | 23.7 | | |
| Cork *(M kg)* | 169 (6.3) | | 140 | | |
| Honey *(M kg)* | 242 (8.9) | | 7.8 | | |
| Soil condition *(C/N)* | | | | | 🟡/🟢 |
| Organic floor | 25.3 | | 31.6 | | |
| mineral 0-10 cm | 17.4 | | 16.4 | | |
| Desertification *(% f. area at risk)* | n/a | | 60 | | 🟡 |
| Forest damage *(% f. area)* | | +2.5% | | | |
| Insects and diseases | 2.8 | 1990–10 | 9.2 | | 🟡 |
| Grazing and wildlife | 2.2 | | 38.0 | | |
| Invasive trees | 0.05 | | 0.17 | | |
| Fire | 0.2 | | 4.2 | | |
| Storms | 1.7 | | n/a | | |

| Variables and indicators | EU-27 | Trend | Portugal | Trend | Evaluation |
|---|---|---|---|---|---|
| Naturalness (*% f. area*) | | | | | |
| Undisturbed by man | 4 | | n/a | | 🟡 |
| Semi-natural | 88 | | 62 | | |
| Plantations | 8 | | 38 | | |
| Deadwood ($m^3\ ha^{-1}$) | 10 | | 1.0 | | 🟡 |
| *(average and range)* | (2–41) | | | | |
| Forest connectivity index *(average normalized connectivity/ landscape unit)* | 0.75 | | 0.7 | | 🟢 |
| Threatened species (*n°*) – 2005 *(average and range)* | | | | | |
| Trees | 6 (0–30) | | 22 | | 🔴 |
| Birds | 30 (1–248) | | 103 | | |
| Mammals | 14 (3–31) | | 19 | | |
| Other vertebrates | 15 (1–81) | | n/a | | |
| Other invertebrates | 97 (4–476) | | n/a | | |
| Vascular plants | 194 (7–1196) | | 144 | | |
| Fungi | 300(77–1284) | | n/a | | |
| Protected forests – Biodiversity C.I & Landscape C.II (*% f. area*) | 18.1 (5–45) | +0.8% 2000–10 | 5.5 | −0.2% | 🟢 |
| Natura 2000 (*% f. area*) | 21.1 | | 23.3 | 1995–2015 | |
| Forest w/management plan (*% f. area*) | 77 (10–100) | | 44 | | 🟡/🟢 |
| Certified forest SFM (*% f. area*) | 62 | | 15.3 | | 🟡/🟢 |
| Contribution GDP (*%*) | 1.0 (0.2–5.0) | | 1.6 | | 🟢/🟡 |
| Export – Import (*M euros*) | 2106 | | 1036 | | 🟢 |

*Reference year 2010 [70–73].* 🟢 *Good/Fair;* 🟡 *Need improvement;* 🔴 *Bad/At risk.*

**Table 1.**
*Status and condition of the forests in Portugal, and comparison with Europe (EU-27).*

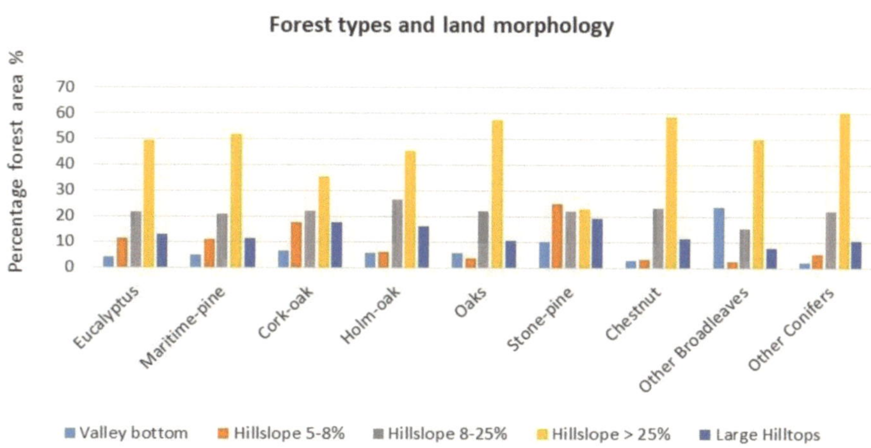


**Figure 7.**
*Different forest types in relation to the land morphology (percentage in relation to each forest type) (collected data using [74]). In Portugal, forest land use covers 3.2 million ha (36.2% of land area) (2015, [71]). The forest and agroforest types and land percentage cover are eucalyptus, mostly* Eucalyptus globulus *Labill. (26.2%), maritime-pine* Pinus pinaster *Aiton. (22.2%), cork-oak* Quercus suber *L. (22.3%), holm-oak* Quercus rotundifolia *Lam. (10.8%), other oaks* Quercus *spp. (2.5%), stone-pine* Pinus pinea *L. (6.0%), chestnut* Castanea sativa *Mill. (1.5%), carob* Ceratonia siliqua *L. (<1%), acacia* Acacia *spp. (<1%), other broadleaves (5.9%) and other conifers (1.6%).*

**Figure 7** shows the relation between each forest type and its location in land morphology, using categories defined by [75]: valley bottoms, hillslopes and large hilltops. The location of the forest in the Portuguese landscape not always fulfills the best soil and water conservation goals. Most forest species and the two species with the higher occupation and use (eucalyptus and maritime-pine plantations) are mostly located in slopes above 25%, which normally represent the less suitability for forestry production, due to the high susceptibility to soil erosion and other issues, considering the silvicultural practices that have been applied.

The study of the dimension of the forest stand also shows that the current landscape has extensive areas of fire-prone eucalyptus and pine plantations and monocultures, with patches over 100 ha (**Figures 8** and **9**) (collected data using [74]). The large amount and contribution to the total forest cover of these forest areas are mainly eucalyptus (18%) and maritime-pine (17%) (**Figure 8**).

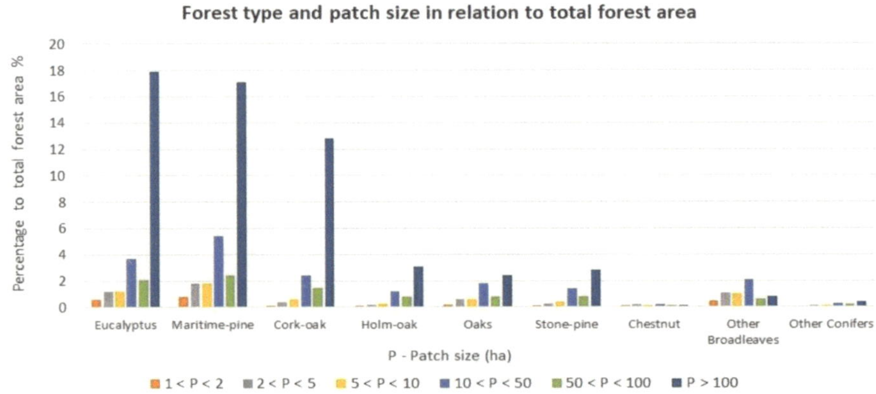


**Figure 8.**
*Forest type and patch size in relation to total forest area in Portugal (contribution of each forest type and patch size to the total country forest area).*

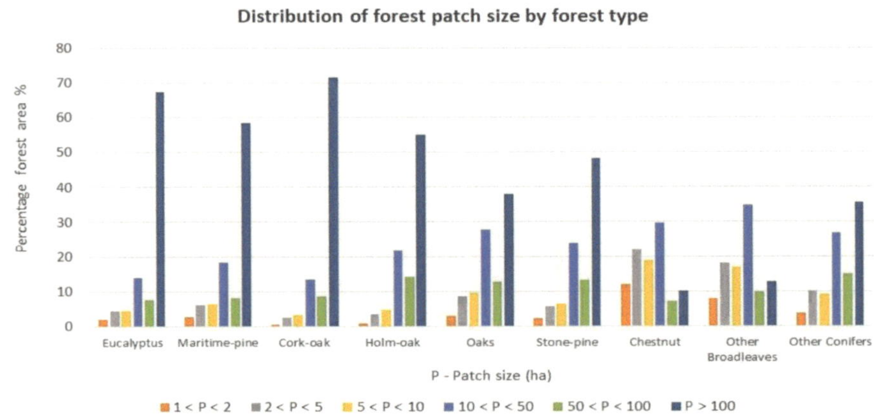


**Figure 9.**
*Distribution of patch size for different forest types (percentage in relation to each forest type).*

## 6.1 The landscape plan

The landscape plan that will define the composition and location of the forest in the landscape must respond to the aspects mentioned: soil conservation, water conservation, biodiversity conservation and conservation of the forest itself, such as many others related to the human activities (accessibility, urban settlements, etc.).

The former text highlights the following keywords: water retention, infiltration, headwater system, land morphology, lithology, permeability, species, combustibility, leaf litter, margins, floodwaters and springs, soil and its texture and structure, building and/or compaction, erosion, cultural practices, continuity, landscape fire-prevention network. All these keywords give an idea of the complexity of the subject of the landscape/land-use planning.

The areas of the river basin where nature conservation issues (*latu sensu*) must be addressed have already been mentioned and constitute the materials of the landscape plan (**Figure 10**). In addition to the location, the species to be used must be defined as well as cultural practices and management models. The species to be used must correspond to their ecological suitability [76], and depending on the role, a specie or mixture of species can perform in the ecosystem [49]. As for the most combustible species, such as *Pinus pinaster* Aiton. and *Eucalyptus globulus* Labill., in case of use, a place may be reserved for them, and always outside areas where nature conservation or ecological restoration is a priority [77], such as extremely degraded areas.

In Portugal, the native species were almost banned, in the name of an economy linked to paper pulp and other wood products, opting for faster growing species managed in monospecific stands that constitute deserts, both animals and people. This model has had severe consequences on the depopulation of the countryside, with emigration and exodus to the cities, living in unhealthy settlements and underpaid jobs, leading to the current high risk of rural fires. It is important to take technical and political measures to improve and valuate native broadleaved species by their several goods and services, combined with management techniques that provide better profitability and contributing to sustainability.

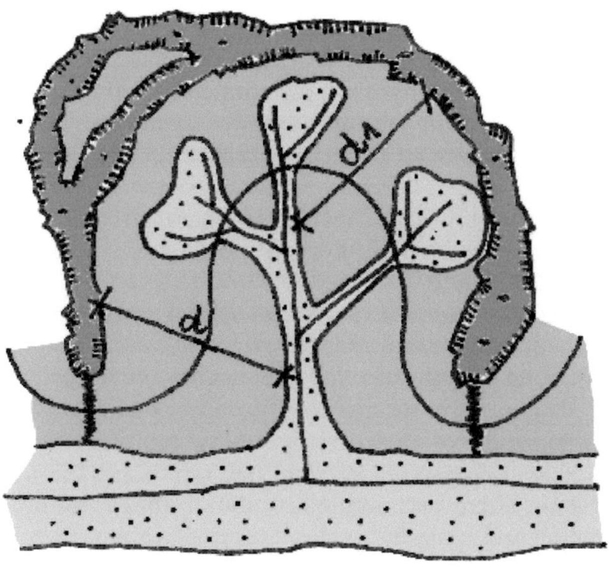


**Figure 10.**
*Conceptual scheme of landscape intervention on a river basin. The valley bottom and streams should be used for agriculture, grazing and/or riparian species; headwater systems should be covered by hardwoods which might be interspersed with void fields intended for grazing purposes. Hillslopes might be covered with woods for production interspersed with longitudinal and transversal autochthonous hardwood species, depending on the slope length (d, d1) and along the contours (adapted from Magalhães et al. [66]).*

It is now necessary to reverse this landscape organization model, through a new paradigm that will take at least three decades to implement and that has to be actively funded by the government.

## 7. Conclusion

The analysis of the forest characteristics, the type of forest and its current location in the landscape of Portugal indicates that the silviculture and the landscape planning paradigm need to be changed.

Forest ecosystems involve biotic and abiotic processes occurring at different spatial and temporal scales, and at different levels of biological hierarchy. A relatively uniform management originates a pattern of relatively low diversity, which results in a loss of some processes and species. An excessive artificiality of the forest also increases their susceptibility to disturbances. Several studies show that complexity improves important forest attributes such as biodiversity, yield, resistance and resilience to several factors or disturbances (e.g., Hansen et al. [78]; Lindenmayer and Franklin [4]). Stand structural diversity, connectivity and landscape heterogeneity are important for ecosystem functioning and biodiversity.

Silviculture may be also related to the spatial and temporal scales in which the different ecological processes occur. It may be incorporated into a planned landscape, which can encompass different ecosystems and forms of intervention, seeking the sustainability of natural resources.

Diversity and landscape interactions can be promoted considering variations on the following main elements: site ecological conditions; composition and structural characteristics of the forest stands; forms of land use (forestry, agro-forestry, agriculture and pasture).

At the landscape level, management requires a spatial and temporal coordination of silviculture applied to different stands, trying the ecological maintenance or restoration of the landscape. At the forest stand level, according to the site conditions, objectives and assigned functions, silviculture suits the stand characteristics related to structure and composition. An integrated ecological-based silviculture provides a set of values that improve the economic efficiency and ecological conditions of the stand. In turn, at the landscape scale, forestry can be combined with other forms of land use, as well as with the variations of the site characteristics that might lead to forest patches with different characteristics and silviculture. In larger forest areas, units can be built differing by their characteristics and objectives, which may allow a diversification of interventions and operations, within a classified division and planning.

Biodiversity should also be evaluated in the spatial and temporal scales. For example, different units are viewed as interacting elements that continuously vary in space and time. Depending on the requirements of a given species, there may be a hierarchical space structuration for different groups, populations and metapopulations. The integration of biodiversity into the multifunctional silviculture is achievable considering a wide living space of habitats where the flora and fauna coexist and interrelate. Therefore, organizing communities and processes requires considering the spatial and temporal scales. Different habitat components can be managed in different scales (tree, stand and landscape). Ideally, various components of biodiversity are considered at different scales. Increasing structural diversity and spatial variability creates different ecological conditions that can promote biodiversity and resilience to disturbances (e.g., Turner et al. [79]).

Silviculture has a major influence on the presence and maintenance of microhabitats. The abundance and diversity of microhabitats considerably increase with tree size and age. The existence of trees with cavities and dead wood is of particular interest to many specialist species. The tree species also plays an important role, and there is interest in certain species by their naturalness, kind of tree-related microhabitats and associated species. Silviculture should allow the maintenance of large trees, dead wood as well as certain species, which exhibit certain features of interest for biodiversity. Trees without economic value may be maintained by its biological value, without decreasing stand production, and avoiding harvesting costs, and, on the other hand, contribute to a better functioning of the ecosystem, with positive effects on the production of the stand.

Besides ecosystems and species diversity, biodiversity also involves genetic diversity. Genetic diversity in a given population of trees is determined by the long evolutionary history and population dynamics. A genetic diversity, in terms of intra- and interpopulation variability, is also an aspect to consider in silviculture given its importance to various levels as element of biodiversity in forest growth and production and survival and adaptability and as a vital part of the ecosystem functioning. This is important in terms of reproductive success, adaptability with implications for evolution, and climate change adaptation.

The tree regeneration is a key process that influences the genetic diversity of the stand population, affecting adaptation and demographic processes. The natural regeneration improves genetic diversity and enables continuous adaptation and evolution of the population in a given location. Evaluations conducted on the application of selective cuts show that natural regeneration has positive effects in genetic diversity [80, 81].

Habitat loss, overexploitation and inadequate silviculture are the main factors of the Mediterranean forest degradation (e.g., Chiatante et al. [82]). Some practices such as clearcutting and fragmentation might have negative effects for soil and water status, as well on the stand regeneration. In turn, an excessive fragmentation can have negative effects on the tree seed predation and dispersion as shown by Santos and Tellería [83] and Morán-López et al. [84].

Reducing fragmentation through appropriate silviculture helps to maintain biodiversity and the ability of forest natural adaptation. The recommended solutions resulting from specific studies vary depending on the biological groups, the forest type, the site characteristics, the distribution pattern and targets [55, 85]. Measures to solve fragmentation should also be combined with actions for forest fire prevention, particularly: actions at the social level; forest partitioning or segregation with other forms of land use (agriculture, agro-forestry and pasture); utilization of forest species more resistant to fire, with a lower combustibility and fire propagation.

Several programs related to the conservation and promotion of biodiversity have highlighted the importance of an integrated approach in silviculture. Some studies have shown that biodiversity conservation involves combining different types of strategies and measures applied at different scales. The level of integration of conservation measures will depend on several factors, namely, related to the ecosystem characteristics, conservation needs and forest management objectives. Several authors (e.g., Kohm and Franklin [86]; Lindenmayer and Franklin [4]) indicate the promotion or maintenance of the following key elements for the biodiversity conservation and its relationship with the landscape: stand structural complexity, connectivity and landscape heterogeneity. Continuity in space and time is an essential element to support populations, since there is a wide range of life strategies and habitat requirements.

Silviculture should enable the development of these elements at the tree, stand and landscape levels. Different values can be achieved or involve different spatial scales. This can create heterogeneity and spatial variability.

## Acknowledgements

The authors would like to thank the support given by the Project SCAPEFIRE (PCIF/MOS/0046/2017) financed by FCT (Science and Technology Foundation). This work was developed in CITAB (UID/AGR/04033/2019) and in LEAF (UID/AGR/04129/2013), financed by the FCT through national funds. The authors also thank Jill Barbour for the text review.

## Author details

João Carvalho[1*], Manuela Magalhães[2] and Selma Pena[2]

1 Department of Forest Science and Landscape Architecture, CITAB, University Tras-os-Montes Alto Douro, Vila Real, Portugal

2 Instituto Superior Agronomia, Linking Landscape Environment, Agriculture and Food (LEAF), University Lisboa, Lisbon, Portugal

*Address all correspondence to: jpfc@utad.pt

## References

[1] Attiwill P. The disturbance of forest ecosystems: The ecological basis for conservative management. Forest Ecology and Management. 1994;**63**: 247-300

[2] Loreau M. Biodiversity and ecosystem functioning: Current knowledge and future challenges. Science. 2001;**294**:804-808

[3] Jax K, Roozi R. Ecological theory and values in the determination of conservation goals. Revista Chilena de Historia Natural. 2004;**77**:349-366

[4] Lindenmayer D, Franklin J. Conserving Forest Biodiversity: A Comprehensive Multiscaled Approach. Washington DC: Island Press; 2002. p. 352

[5] Harrison S, Fahrig L. Landscape pattern and population conservation. In: Hansson L, Fahrig L, Merriam G, editors. Mosaic Landscapes and Ecological Processes. London: Chapman & Hall; 1995. pp. 293-308

[6] Turner M, Gardner R, O'Neill R. Ecological dynamics at broad scales. Bioscience. 1995;**45**:29-35

[7] West D, Shugart H, Botkin D. Forest Succession. NY: Springer-Verlag; 1981

[8] Tilman D. Plant Strategies and the Dynamics and Structure of Plant Communities. Princeton NJ: Princeton Univ. Press; 1988. p. 376

[9] Oliver C, Larson B. Forests Stand Dynamics. NY: McGraw-Hill; 1990. p. 467

[10] Grime J. Plant Strategies, Vegetation Processes and Ecosystem Properties. Bafins Lane: John Wiley & Sons; 2002. p. 417

[11] Hooper D, Solan M, Symstad A, Diaz S, Gessner M, Buchmann N, et al. Species diversity, functional diversity and ecosystem functioning. In: Loreau M et al., editors. Biodiversity and Ecosystem Functioning - Synth. Perspect. Oxford Uni. Press; 2002. pp. 195-208

[12] Jax K. Ecosystem Functioning. Cambridge Univ Press; 2010. p. 272

[13] Bouget C, Larrieu L, Nusillard B, Parmain G. In search of the best local habitat drivers for saproxylic beetle diversity in temperate deciduous forests. Biodiversity and Conservation. 2013;**22**:2111-2130

[14] White P. Pattern, process and natural disturbance in vegetation. The Botanical Review. 1979;**45**:229-299

[15] White P, Pickett S. Natural disturbance and patch dynamics. In: Pickett S, White P, editors. The ecology of natural disturbance and patch dynamics. NY: Academic Inc.; 1985. p. 121

[16] Bazzaz F. Characteristics of population in relation to disturbance in natural and man-modified ecosystems. In: Money H, Godron M, editors. Disturbance and Ecosystems: Components of Response. NY: Springer-Verlag; 1983. pp. 259-275

[17] Pickett S, McDonnell M. Changing perspectives in community dynamics: A theory of successional forces. Trees. 1989;**4**:241-245

[18] Hunter J. Benchmarks for managing ecosystems: Are human activities natural? Conservation Biology. 1996;**10**: 695-697

[19] Crooks KR, Sanjayan M. Connectivity conservation: Maintaining connections for nature. In: Crooks KR, Sanjayan M, editors. Connectivity Conservation. Cambridge: Cambridge University Press; 2006. pp. 1-19

[20] Ellenberg H. Vegetation Ecology of Central Europe. Cambridge Univ. Press; 1988. p. 731

[21] Clements F. Plant succession. In: An analysis of the Development of Vegetation. Washington DC: Carnegie Inst.; 1916. p. 658

[22] Gleason H. The individualistic concept of the plant association. Bulletin of the Torrey Botanical Club. 1926;**46**: 7-26

[23] Noble I, Slayter R. The use of vital attributes to predict successional changes in plant communities subject to recurrent disturbance. Vegetatio. 1980; **43**:5-21

[24] Whittaker R. Climax concepts and recognition. Handbook of Vegetable Science and Technology. 1973;**8**:137-154

[25] Mueller-Dombois D, Ellemberg H. Aims and Methods of Vegetation Ecology. NY: John Wiley & Sons; 1974. p. 545

[26] Pickett S, White P. The Ecology of Natural Disturbances and Patch Dynamics. Orlando: Academic Press; 1985. p. 472

[27] Swanson F, Kratz T, Caine N, Woodmansee R. Landform effects on ecosystem patterns and processes. Bioscience. 1988;**38**:92-98

[28] Pickett S, Ostfeld R. The shifting paradigm in ecology. In: Knight R, Bates S, editors. A New Century for Natural Resources Management. Washington DC: Island Press; 1995. pp. 261-278

[29] McGill B, Enquist B, Weiher E, Westoby M. Rebuilding community ecology from functional traits. Trends in Ecology & Evolution. 2006;**21**:178-185

[30] Wiens J. Spatial scaling in ecology. Functional Ecology. 1989;**3**:385-397

[31] Colinvaux P. Initiation à la science de l'écologie. Ed Seuil, Paris: Points Sciences; 1993

[32] McCann K, Hastings A, Huxel G. Weak trophic interactions and the balance of nature. Nature. 1998;**395**: 794-798

[33] Neutel A, Heesterbeek J, van de Koppel J, Hoenderboom G, Vos A, Kaldeway C, et al. Reconciling complexity with stability in naturally assembling food webs. Nature. 2007; **449**:599-602

[34] Spies T. Forest stand structure, composition and function. In: Kohm K, Franklin J, editors. Creating a Forestry for the 21st Century. Washington DC: Island Press; 1997. pp. 11-30

[35] Bormann F, Likens G. Pattern and Process in a Forested Ecosystem. NY: Springer-Verlag; 1994. p. 253

[36] Canham C, Marks P. The response of woody plants to disturbance. In: Pickett S, White P, editors. The Ecology of Natural Disturbances and Patch Dynamics. NY: Academic; 1985. pp. 197-216

[37] Hibbert R. Forest treatment effects on water yield. In: Sopper W, Lull H, editors. Intern Symposium Forest Hydrology. NY: Pergamon Press; 1976. pp. 527-543

[38] Diochon A, Kellman L, Beltrami H. Looking deeper: An investigation of soil carbon losses following harvesting from a managed northeastern red spruce (Picea rubens Sarg.) forest chronosequence. Forest Ecology and Management. 2009;**257**:413-420

[39] Covington W. Changes in forest floor organic matter and nutrient content following clear cutting in northern hardwood. Ecology. 1981;**62**: 41-48

[40] Covington W, Aber J. Leaf production during secondary succession in northern hardwoods. Ecology. 1980; **61**:200-204

[41] Spies T, Franklin J, Thomas T. Coarse wood debris in Douglas-fir forests of western Oregon and Washington. Ecology. 1988;**69**: 1689-1702

[42] With K, Gardner R, Turner M. Landscape connectivity and population distributions in heterogeneous environments. Oikos. 1997;**78**:151-169

[43] Cunha NS, Magalhães MR. Methodology for mapping the national ecological network to mainland Portugal: A planning tool towards a green infrastructure. Ecological Indicators. 2019;**104C**:802-818

[44] Corte Real J. Não estamos à beira de qualquer catástrofe. Expresso; 2008

[45] Magalhães MR. A Arquitectura Paisagista Morfologia e Complexidade. Lisbon: Editorial Estampa; 2001. p. 525

[46] Magalhães M. La Struttura Ecologica Metropolitana di Lisbona. In: Marano S, editor. Ridurre lo Spazio. Gangemi Editore, Roma: Frammento e Specificità dei Luoghi; 2006. pp. 86-92

[47] Pena S, Abreu M, Magalhães M. Planning landscape with water infiltration. Empirical model to assess maximum infiltration areas in Mediterranean Landscapes. Water Resources Management. 2016;**30**: 2343-2360

[48] Pena S, Magalhães M, Abreu M. Mapping headwater systems using a HS-GIS model. An application to landscape structure and land use planning in Portugal. Land Use Policy. 2018;**71**: 543-553

[49] Molchanov A. Hidrologia Florestal. Lisbon: Fund Calouste Gulbenkian; 1971

[50] Pena S, Abreu M, Magalhães M, Cortez N. Water erosion aspects of land degradation neutrality to landscape planning tools at national scale. Geoderma. 2020;**363**:114093

[51] Harris L. The Fragmented Forest. Chicago Press: The Univ; 1984. p. 211

[52] Tilman D, Lehman C, Yin C. Habitat destruction, dispersal, and deterministic extinction in competitive communities. The American Naturalist. 1997;**149**: 407-435

[53] Collingham Y, Huntley B. Impacts of habitat fragmentation and patch size upon migration rates. Ecological Applications. 2000;**10**:131-144

[54] Tellería J, Díaz J, Pérez-Trís J, Santos T. Fragmentación de hábitat y biodiversidade en las mesetas ibéricas: una perspectiva a largo plazo. Ecosistemas: Revista Cietifica y Tecnica de Ecologia y Medio Ambiente. 2011;**20**: 79-91

[55] Tambosi L, Martensen A, Ribeiro M, Metzger J. A framework to optimize biodiversity restoration efforts based on habitat amount and landscape connectivity. Restoration Ecology. 2014; **22**:169-177

[56] Swanson F, Franklin J. New forestry principles from ecosystem analysis of Pacific Northwest Forests. Ecological Applications. 1992;**2**:262-274

[57] Tilman D, Kareiva P. The role of space in population dynamics and interspecific interactions. In: Population Biology 30. New Jersey: Princeton University Press; 1997. p. 368

[58] Young A, Boyle T, Brown T. The population genetic consequences of habitat fragmentation for plants. Trends in Ecology & Evolution. 1996;**11**:413-418

[59] Rothermel R. How To Predict the Spread and Intensity of Forest and

Range Fires. Ogden: Inter-mountain. Forest and Range Exp Stat; 1983. p. 164

[60] Moreira F, Duarte I, Catry F, Acácio V. Cork extraction as a key factor determining post-fire cork oak survival in a mountain region of southern Portugal. Forest Ecology and Management. 2007;**253**:30-37

[61] Heyerdahl E, Brubaker L, Agee J. Spatial controls of historical fire regimes: A multiscale example from interior west. Ecology. 2010;**82**(3): 660-678

[62] Oliveira S, Moreira F, Boca R, San-Miguel-Ayanz J, Pereira JMC. Assessment of fire selectivity in relation to land cover and topography: A comparison between Southern European countries. International Journal of Wildland Fire. 2014;**23**(5): 620-630

[63] McArthur AG. Fire Behaviour in Eucalypt Forests. India: Leaflet 107. Ninth Commonwealth Forestry Conference; 1968

[64] Agee JK, Bahro B, Finney MA, Omi PN, Sapsis DB, Skinner CN, et al. The use of shaded fuelbreaks in landscape fire management. Forest Ecology and Management. 2000;**127** (1–3):55-66

[65] Povak NA, Hessburg PF, Salter RB. Evidence for scale-dependent topographic controls on wildfire spread. Ecosphere. 2018;**9**(10):e02443

[66] Magalhães MR, Batista FO, Cunha N, Müller A, Pena SB, Silva J, et al. O Ordenamento do Território na Prevenção dos Incêndios Rurais. In: Tedim F, Paton D, editors. A dimensão humana dos incêndios florestais. Porto: Estratégias Criativas; 2012. pp. 55-99

[67] Silva JS, Moreira F, Vaz P, Catry F, Godinho-Ferreira P. Assessing the relative fire proneness of different forest types in Portugal. Plant Biosystems. 2009;**143**(3):597-608

[68] Calviño-Cancela M, Chas-Amil M, García-Martínez E, Touza J. Interacting effects of topography, vegetation, human activities and wildland-urban interfaces on wildfire ignition risk. Forest Ecology and Management. 2017; **397**:10-17

[69] Massari G. Leopaldi, A. Leaf flammability in mediterranean species. Plant Biosystems. 2013;**132**(1):29-38

[70] ICNF. Inventário Florestal Nacional IFN5. Lisbon: ICNF; 2010

[71] ICNF. Inventário Florestal Nacional IFN6. Lisbon: ICNF; 2015

[72] MCPFE. State of Europe's Forests. Oslo: Liaison Unit; 2011

[73] INE. Estatísticas da Produção Industrial, 2010. Lisbon: INA; 2011

[74] DGT. Carta Uso Ocupação do Solo Portugal Continental 2018. Lisbon: DGT; 2019

[75] Cunha N, Magalhães M, Domingos T, Abreu M, Withing K. The land morphology concept and mapping method and its application to mainland Portugal. Geoderma. 2018;**325**:72-89

[76] Mesquita S, Capelo J. Aptidão bioclimática às espécies arbóreas. In: Magalhães. Lisbon: M. Ordem Ecológica e Desenvolvimento. O futuro do território português. ISAPRESS; 2016. pp. 63-83

[77] Schirone B, Salis A, Vessela F. Effectiveness of the Miyawaki method in Mediterranean forest restoration programs. Landscape and Ecological Engineering. 2011;**7**(1):81-92

[78] Hansen A, Spies T, Swanson F, Ohmann J. Conserving biodiversity in

managed forests. Bioscience. 1991;**41**: 382-392

[79] Turner M, Donato D, Romme W. Consequences of spatial heterogeneity for ecosystem services in changing forest landscapes: Priorities for future research. Landscape Ecology. 2012;**28**: 1081-1097

[80] Raja R, Tauer C, Wittwer R, Huang Y. Regeneration methods affect genetic variation and structure in shortleaf pine (Pinus echinata Mill.). Forest Genetics. 1998;**5**:171-178

[81] Westergren M, Bozic G, Ferreira A, Kraigher H. Insignificant effect of management using irregular shelterwood system on the genetic diversity of European beech (Fagus sylvatica L.): A case study of managed stand and old growth forest in Slovenia. Forest Ecology and Management. 2015; **335**:51-59

[82] Chiatante D, Domina G, Montagnoli A, Raimondo F. Sustainable restoration of Mediterranean forests. Flor Medit. 2017;**27**:5-76

[83] Santos T, Tellería JL. Vertebrate predation on Holm Oak, Quercus ilex, acorns in a fragmented habitat: Effects on seedling recruitment. Forest Ecology and Management. 1997;**98**:181-187

[84] Morán-López T, Fernández M, Alonso C, Flores-Rentería D, Valladares F, Díaz M. Effects of forest fragmentation on the oak-rodent mutualism. Oikos. 2015;**124**:1482-1149

[85] Franklin J. Preserving biodiversity: Species, ecosystems, or landscapes. Ecological Applications. 1993;**3**:202-205

[86] Kohm K, Franklin J, editors. Creating a Forestry for the 21st Century. Washington DC: Island Press; 1997. p. 475

Chapter 7

# Ecological and Social Impacts of Aquacultural Introduction to Philippines Waters of Pacific Whiteleg Shrimp *Penaeus vannamei*

*Marlon S. Alejos, Augusto E. Serrano, Jr., Yashier U. Jumah, Rey dela Calzada, Cyril Tom B. Ranara and Jumari C. Fernandez*

## Abstract

The importation of exotic aquatic species for aquaculture has become a worldwide practice. Culturing the Pacific whiteleg shrimp, *Penaeus vannamei*, outside its natural geographic range, however, can cause many problems. We evaluated the implications of the aquacultural introduction of *P. vannamei* to the Philippines waters both on ecological and social aspects. Several questions were answered and discussed based on literature, scientific details, reflections on personal experience and their relevance to aquaculture of the *P. vannamei* in the Philippines to evaluate the ecological impacts while social impacts were discussed only based on literature. Findings revealed the escapes of *P. vannamei* from aquaculture production facilities of several countries including the Philippines. Consequently, the ability of *P. vannamei* escapees to survive the natural environment could lead to ecological concerns such as resource competition, reproduction, and the spread of disease in the wild. On the other hand, the recent expansion of shrimp culture has resulted in social conflicts with other resource users. Therefore, this review shows the negative implications on the aquacultural introduction of Pacific whiteleg shrimp *P. vannamei* in the Philippines both ecological and social aspect, and this heightens important management issues to ensure sustainable farming of the shrimp in the Philippines.

**Keywords:** exotic organism, intentional release, tolerance, propagule pressure, alien pathogen

## 1. Introduction

An exotic organism or a non-native organism is a plant or animal that has been transplanted by humans; they are usually perceived from a negative point of view [1]. The International Union for Conservation of Nature [2] describes Alien species (non-native, non-indigenous, foreign, exotic) a species, subspecies, or lower taxon occurring outside of its natural range (past or present) and dispersal potential (i.e., outside the range it occupies naturally or could not occupy without direct or indirect introduction or care by humans), which includes any part, gametes, or propagule of such species that might survive and subsequently reproduce. On the other hand, Invasive alien species (IAS) are species whose introduction and/or spread outside their natural past or present distribution threatens biological diversity which occurs in all taxonomic groups, including animals, plants, fungi, and microorganisms, and can affect all types of ecosystems [3]. Generally, exotic species are regarded as IAS.

IAS is one of the five most important direct drivers of biodiversity loss and change in ecosystem services [4]. According to the Invasive Special Specialist Group, [5], IAS can interact with migratory species in several ways resulting in cumulative negative impacts, for example, as a threat on their breeding sites, on their stopover and wintering grounds, and during migrations. These impacts may result in local extinction or decline in population numbers as well as changes to migration patterns. IAS impacts native species (including migratory species) and their habitats through several mechanisms, including predation, habitat degradation (grazing, herbivory, browsing, rooting/digging and trampling), competition, hybridization, disease transmission, parasitism, poisoning/toxicity, biofouling, etc. IAS has resulted in major impacts on biodiversity at a global scale, where at least 39% of the species extinctions during the past 400 years are due to IAS [6].

The importation of alien or exotic aquatic species from other countries is continuing in the Philippines. Most of the importation is for aquaculture and the aquarium trade. Exotic species are either purposely or accidentally introduced in rivers and lakes which are inhabited by endemic and indigenous fish species. Most of these introductions have contributed negative impacts on freshwater/wetland ecosystems and have caused biodiversity loss [7, 8], while some introduced species have contributed a significant proportion to aquaculture in the Philippines [9]. In terms of such contributions, the Nile tilapia *Oreochromis niloticus* is next to milkfish *Chanos chanos* among the aquaculture species, followed by the big head carp *Aristichthys nobilis* [10].

There are 181 organisms (28 families) introduced of exotic aquatic species since the 1900s; however, 40 organisms have unknown records of introductions in the Philippines [11]. In 2018, the IUCN Invasive Species Specialist Group [12]'s Global Invasive Species Database (GISD) lists 84 alien species, 12 with bio status unspecified and 54 that are native to the Philippines. The Pacific whiteleg shrimp *P. vannamei* is not included in the list; however, published studies reported the risk of culturing the shrimp outside its natural geographic range. For intensively farmed *P. vannamei* in Indonesia, the final numerical score is 3.39 out of 10, where the presence of three red criteria (Habitat, Chemicals, Disease) results in an overall red "Avoid" recommendation. Red mark means that these items are overfished or caught or farmed in ways that harm other marine life or the environment [13].

*P. vannamei* [14] is native to the Eastern Pacific coast from the Gulf of California, Mexico to Tumbes, North of Peru [15]. In the Philippines, *P. vannamei* was imported from Panama in the 1970s and from Hawaii in 1990 [16, 17]. However, there was no documentation of these introductions because government regulations were not followed in most cases, and it was not known whether the exotic species introduced any new pathogens [18].

Risk can be defined as the likelihood of harm occurring as a result of an action or inaction [19]. Harm refers to the undesirable consequences to humans or components of a valued ecosystem [20]. However, Senanan et al. [21] argued that there is a challenge in analyzing the ecological risk of alien species because of difficulty in predicting the harm, estimating the likelihood of harm occurring and the severity of the harm. These parameters are species and ecosystem specific and are often difficult to measure. The analysis would describe the ecosystem components (abiotic and biotic) and relevant processes, life history characteristics of the alien species and the relevant interaction between the two, drawing on existing data and literature, as well as specific experiments and field surveys.

The presence of Pacific whiteleg shrimp *P. vannamei* in the Philippines has already reached more than five decades, and this raises risk and environmental concern. This is timely relevant to the theme of the book *Spatial Variability in the Environmental Sciences - Patterns, Processes, and Analyses* which covers the topics on migration, extinction, disturbance, restoration, contamination, conservation, pollution, revitalization, growth, and decline. Thus, this review paper aimed to determine the ecological and social impacts of aquacultural introduction to the Philippines waters of Pacific whiteleg shrimp *P. vannamei*. The preliminary risk analysis of the releases of *P. vannamei* guided with several questions reported by Senanan et al. [22] was adopted to evaluate the ecological impacts, while social impacts were evaluated based on literature. Guide questions include (1) "How many *P. vannamei* have escaped?", (2) "Can escapees survive in the natural environment?", (3) "Can escapees establish a natural population?", (4) "What is the extent of the geographic spread of the alien pathogen, Taura syndrome virus (TSV)?", and (5) "Can *P. vannamei* potentially compete with native shrimp species?". The questions were answered and discussed based on literature, scientific details, reflections on personal experience, and their relevance to aquaculture of the *P. vannamei* in the Philippines. Conclusions were formulated based on the interpreted findings of this review, and recommendations were made for sustainable aquaculture of *P. vannamei* in the Philippines.

## 2. How many *P. vannamei* have escaped?

There have been numerous reports of escapes from aquaculture production facilities into non-native waters. The presence of *P. vannamei* has been reported in Texas, South Carolina, and Hawaii, USA [23–28]; Thailand [29, 30]; Venezuela [31]; Brazil [32]; Puerto Rico [33]; Vietnam [34]; and Southern Gulf of Mexico coast [35]. In the Philippines, Briggs et al. [29] reported that a population of *P. vannamei* already exists in the wild through intentional release and escapes. The implementation to ban the importation of all live shrimp and prawn species of all stages except for scientific or educational purposes by the Bureau of Fisheries and Aquatic Resources (BFAR) in 1993 led to illegal importation in 1997 by private sector due to disease problems with the culture of *P. monodon*, and the regulations are known to have resulted in the dumping of PL *P. vannamei* into the wild in attempts to escape detection. Also, typhoons have also resulted in the liberation of *P. vannamei* from culture ponds into the surrounding sea. On average, 20 typhoons hit the Philippines every year, and some of the most destructive and deadliest typhoons include Yolanda (2013), Pablo (2012), Sendong (2011), Ondoy (2009), Frank (2008), Milenyo (2006), and Reming (2006) (http://bagong.pagasa.dost.gov.ph/). Based on the main author personal experience, *P. vannamei* has been sometimes a part of catch by local fisherman in Buguey Lagoon, Cagayan. Its presence in the lagoon is possibly due to the escape from the Dataj Aquafarm which is actively engaging in the grow-out of *P. vannamei* in four different locations in the municipality of Buguey and Camalaniugan with a total area 77.91 hectares. The lagoon serves as the main water source of the farms. More frequent flood incidence in the area is taking place especially during typhoons or heavy rains due to black sand mining that started in 2009 until 2013 that widens the mouth of the lagoon.

Through the years, the culture of *P. vannamei* in the Philippines is continuously growing even during the implementation of the shrimp importation ban from 1993 to 2006. In fact, about 700 hectares of *P. vannamei* illegal farms in Luzon have been reported in 2003 [36]. The ban was lifted in 2007 after experimental trials and a series of public consultation and hearings [37]. In 2008, there were 38 grow-out farms engaged in *P. vannamei* farming, and it increased to 53 farms in 2013 [38]. Additionally, based on the [39] master list of shrimp farms as of December 2013, the shrimp has been polycultures with fish (17 farms, 642.7 ha) and *P. monodon* (4 farms, 33.0 ha). In 2014, there were 27 accredited hatcheries that continually support the demand of seedling requirements of the shrimp [40]. As of August 31, 2019, however, there were already 40 accredited hatcheries and 545 grow-out farms of the shrimp with a total productive area of 7382.08 hectares [41–43]. The boom and rapid expansion of the shrimp aquaculture are expected because of its demand for both the local and global markets. With this continuous expansion of *P. vannamei* farming, it is possible that shrimp can be found in the different estuary or brackish water rivers in the Philippines wherein the hatcheries and grow-out farm's operations are found due to escapes. The probability of escapes is higher because out of the total farms in 2019, 331 farms with an area of 4144.19 hectares were operating extensive grow-out practices wherein old and not properly designed ponds are usually utilized. A small percentage of escapes per operation cycle could translate to significant numbers of individuals entering the ecosystem. According to Panutrakul et al. [44], the shrimp can enter estuary at various life stages including post-larvae (produced from hatcheries), juveniles, and subadults (cultured in ponds). As of 2019, however, no studies have been conducted in the Philippines to verify the presence of the shrimp in the wild.

In Thailand, floods in Surat Thani and Pranburi in 2003, for example, led to several million *P. vannamei* escaping to the coastal environment. Not surprisingly, the shrimp, therefore, has been reported in fisherfolk's catches on Andaman and Gulf of Thailand coasts. No detailed information on catches is available, but numbers have not been reported as large [29]. Similarly, an incident of escape was reported from farms to the Bangpakong River in Thailand [30, 45]. The study showed that the numbers of the shrimp sampled in the river positively correlated with the location and area of shrimp ponds. Manthachitra et al. [45] used remote sensing and a geographic information system (GIS) to estimate the location and total area of shrimp ponds (active, inactive, and abandoned ponds) in the Bangpakong River watershed and found that most ponds were located within 5 km of the river. The presence of the shrimp was confirmed based on the survey of marine shrimp populations in the Bangpakong River during the same period conducted by Senanan et al. [30]. The mean proportion of *P. vannamei* relative to all penaeid shrimp per net per year (all stations combined) ranged from 0.005 (June 2005) to 0.16 (January 2006), with the highest abundance detected in 2006. The presence of the shrimp in the river may be a consequence of pond water releases during the intense farming activities of 2005. In the Southern Gulf of Mexico coast, there is a first report of the presence of *P. vannamei* [35]. During a shrimp monitoring program survey conducted in this area, seven specimens were collected in the Carmen-Pajonal-Machona lagoons near La Azucena and Sanchez Magallanes in Tabasco, Mexico.

## 3. Can escapees survive the natural environmental conditions?

Pacific white shrimp *P. vannamei* is the most economically important species for aquaculture in extensive, semi-intensive, and intensive systems in many parts of the world [46, 47] due to several advantages compared with other cultured shrimps. These are largely associated with the ability to close the life cycle and produce broodstock within the culture ponds, rapid growth rate (at up to 3 g/week), tolerance of high stocking density (150/m$^2$ in pond culture, and even as high as 400/m$^2$ in controlled recirculated tank culture), tolerance of low salinities (0.5–45 ppt) and temperatures (15–33°C), lower protein requirements (20–35%) and therefore production costs, and high survival (50–60%) during larval rearing [29]. However, the question remains if they can survive in the natural environment wherein adverse conditions are present.

It can be assumed that *P. vannamei* escapees can survive the natural environmental conditions based from early reports [23–31, 34, 35]. For instance, a large number were released accidentally from a shrimp farm in Texas in 1991, and the escapees were caught up to 65 miles from the shore [48]. The presence of the shrimp in commercial catches in South Carolina was also reported in 1989 and 1990 [49]. On the other hand, Medina-Reyna [50] reported the growth and emigration of the shrimp in the Mar Muerto Lagoon, which is one of the largest nursing grounds for this species in Mexico. Reports were all related to the ability of the shrimp to tolerate a wide range of salinity. Recently, the study of Chavanich et al. [51] results indicated that *P. vannamei* escapees can likely survive the environmental conditions of the Bangpakong River and its river mouth. A toxicological experiment was conducted to evaluate the physiological limits of larvae and juveniles of *P. vannamei* and *P. monodon* to extreme salinity and pH changes [44]. Results showed that both species can tolerate a wide range of salinity and pH. For both life stages, *P. vannamei* could tolerate a wider range and more extreme changes of salinity and pH than *P. monodon*. The data suggested that both life stages of *P. vannamei* could adapt to estuarine conditions of the Bangpakong River where water quality, especially salinity, can fluctuate dramatically. The shrimp is also capable of migrating to the river mouth; in times the Bangpakong River may approach zero salinity at most sites during the wet season (June to November). This eventually resulted in an increase in abundance and size overtime of the shrimp captured in the river and near the river mouth. More likely, this scenario already existed in Buguey Lagoon in Cagayan since the shrimp has been a part of fishermen catch in the area, and this could be true in other estuaries and rivers in the Philippines wherein the shrimp hatcheries and farms are located.

## 4. Can escapees establish a natural population?

According to CABI [52], the species itself is not considered a major threat to biodiversity and does not appear to have formed breeding populations. Briggs et al. [29] added that despite the fact that the species has been widely introduced, a comprehensive study of the literature carried out for this report and the information available from other countries in Asia and in the Americas did not find any evidence of the shrimp becoming established in the wild outside of its range (i.e., it may not become an easily "invasive" species). However, there is a need for further field research, as there was insufficient information available on the natural breeding habits of the shrimp to make any further assessment of this issue.

Understanding the biology of *P. vannamei* is vital in order to know whether this animal once escape can establish a natural population that is crucial to the environment. However, there is limited information on the biology of this animal and is neglected during the last two decades. According to Dugassa and Gaetan [53], *P. vannamei* lives in tropical marine habitats, and the adults of this species live and spawn in the ocean. However, the larvae and juveniles are usually found in inshore water areas such as coastal estuaries, lagoons, or mangrove areas. The shrimp females grow faster than the male of this species. The matured female weighing 30–45 g can spawn 100,000–250,000 eggs. The shrimp life cycle is very complex, and it usually takes around 1.5 years to complete the whole life cycle. The matured shrimp females spawn their eggs in the offshore waters [54], while fertilization occurs in the external environment [53]. However, the maturity of the shrimp escaped from farms to natural environments is an important factor in determining their ability to establish a feral population [22]. A study has been conducted to compare the histology of gonads of wild-caught and captive *P. vannamei* of known ages [55]. Captive individuals could develop mature gonads at 11 months after post-larvae 15 (ovaries contained 50% mature oocytes; testes contained 80% mature sperm cells). The result of the study showed that they did not find sexually mature individuals in the wild although some wild-caught males larger than 19 g contained a small percentage of mature sperm cells. However, the authors cannot conclude that escapees can establish a feral population because the study might have under-sampled sexually mature individuals due to inappropriate sampling sites and timing. This issue remains important for further investigation. A monitoring program in offshore areas may provide opportunities for us to obtain sexually mature individuals. Likewise, Panutrakul et al. [44] found no evidence that the shrimp present in the wild could reach maturation in the Bangpakong River although gonadal development has been observed. Moreover, Wakida-Kusunoki et al. [35] argued that it was not possible to find evidence of *P. vannamei* becoming established in the zone of the Mexican coast of the Gulf of Mexico. The low frequency of *P. vannamei* encounters in the monitoring program of artisanal shrimp fishing in lagoon system and the negative presence of the shrimp in surveys of the commercial shrimp catches of coastal waters near to the mouth of this lagoon indicate the absence of an established population. They suggested further sampling and monitoring are required to find evidence that confirms the establishment of a population of the shrimp in the Southern Gulf of Mexico. According to Panutrakul et al. [44], natural reproduction would require released adults and a high probability that the mature adults could find mates. In the Philippines, the population of *P. vannamei* in the wild therefore already exists, although it is still uncertain if this population is now breeding [29] until this time.

## 5. Can *P. vannamei* potentially compete with native shrimp species?

Studies have shown that there is a potential risk of a negative impact of the introduced Pacific whiteleg shrimp *P. vannamei* on native species and the invaded ecosystems [21, 30, 44]. An alien species like *P. vannamei* could potentially interact with local species through food competition, either by exploitative or interference competition [21].

Recently, Chavanich et al. [56] conducted a laboratory assessment of feeding-behavior interactions between the introduced *P. vannamei* and five native shrimps plus a crab species in Thailand. Results showed that the shrimp was nonselective with respect to the palatability of the five native shrimps as food. The shrimp was behaviorally dominant when competing for food one-on-one with the native shrimp species. According to Gamboa-Delgado et al. [57], the shrimp is an opportunistic feeder that can adapt well to changes in diet composition. Though laboratory studies could not represent the feeding interactions under field conditions, the non-native shrimp could become a serious threat to native shrimps when the frequency of escapes is increasing and when they begin to reproduce successfully. One of the key factors influencing the success of invading species is propagule pressure or total quantity [58, 59]. Increasing the propagule pressure may enhance the foundation of an invasive population [58]. In Bangpakong estuary, increased frequency of encountering the shrimp is reflecting an increase in propagule pressure because the frequency of escapes is increasing [30]. However, Chavanich et al. [56] suggested more studies are needed to provide insights into the interactions between the introduced white shrimp and native shrimp species and into the ecosystem-wide consequences of this introduction.

## 6. What is the extent of the geographic spread of the alien pathogen, TSV?

Diseases are worldwide top issues and challenges in shrimp aquaculture based on the survey of global aquaculture alliance from 2016 to 2017 [60]. Modern shrimp farming is, in a way, shaped by viral disease outbreaks in the nineties and early 2000s [61]. On the other hand, Itsathitphaisarn et al. [62] reiterated that viral pathogens pose a primary threat to global shrimp aquaculture. According to Lightner and Redman [63], there are about 20 viral pathogens that can cause serious epizootics in penaeid shrimp. In the Philippines, major viral pathogens affecting the shrimp aquaculture include white spot disorder infection (WSSV), monodon baculovirus (MBV), irresistible hypodermal and hematopoietic rot infection (IHHNV), hepatopancreatic parvovirus (HPV), yellow head infection (YHV), and Taura disorder infection (TSV) [61].

A disease caused by TSV was first described from Ecuador in the early 1990s. Lightner [64] and Lightner [65] reported that the disease outbreaks caused catastrophic losses with cumulative mortality rates of 60 to >90% in pond-cultured shrimp. The principal host species for TSV are the *P. vannamei* and the *P. stylirostris*, and it has been documented in all life stages (i.e., post-larvae, juvenile, adults) of *P. vannamei* except in eggs, zygote, and larvae [66]. TSV is a particularly virulent pathogen of *P. vannamei*, and it can infect several other shrimp species including *P. monodon*, *P. aztecus*, *P. duorarum*, *Litopenaeus setiferus*, *L. stylirostris*, *Marsupenaeus jaiponicus*, *Macrobrachium rosenbergii*, *Metapenaeus ensis*, *Fenneropenaeus chinensis*, and *L. schmitti* [61].

According to Wertheim et al. [67], TSV is now widely distributed in the shrimp-farming regions of the Americas, Southeast Asia, and the Middle East. Additionally, evidence showed that TSV is present in natural populations of *P. vannamei* in Central America such as Mexico and Ecuador and may be elsewhere [66, 68]. The international trade of live shrimp resulted in a rapid spread of TSV in the Americas and Asia [69]. TSV was introduced to Asia in 1998 by careless importation of shrimp stocks for aquaculture but has not been reported to cause problems with local crustacean species [70]. Recently, Thitamadee et al. [71] reported TSV has become innocuous due to the widespread use of highly tolerant specific-pathogen-free (SPF) stocks of *P. vannamei* that dominate production.

In the Philippines, there was no documentation on the introduction of *P. vannamei* from Panama into Iloilo in the 1970s and from Hawaii in 1990, and it was not known whether the exotic species introduced any new pathogens [18]. As of 2015, there was no documented report regarding TSV presence in the Philippines [61] and have yet to be detected as stated in the NACA, OIE, and FAO [72] quarterly animal disease report of 2018. However, this will not justify that TSV is not present in the Philippines because there was no study conducted on the detection of the viral disease since the introduction of *P. vannamei* until the importation ban was lifted in 2007 up to 2018. Rosario and Lopez [36] reported that even with the strict implementation of the BFAR formulated FAO 207 series of 2001 which further strengthened FAO 189 series of 1993 which among others prohibit the importation of exotic shrimps and strict surveillance in airports, traders were finding other ways in bringing the illegal shrimp inside the Philippines without passing through the airports. Shrimp Importation, Monitoring, and Surveillance (SIMS) team spearheaded six major confiscations in late 2002 up to 2003 and reported around 700 have illegal *P. vannamei* farms operating in Luzon. Moreover, despite all the efforts of the BFAR, the culture industry for the shrimp in the Philippines has grown and may produce as much as 5000 metric tons in 2003 [29]. In fact, Philippines was one of the main producer countries of *P. vannamei* based on FAO fishery statistics in 2006 [73]. According to de la Peña [74], there is always the possibility of contamination with TSV if the illegal shipments of the shrimp remain uncontrolled, and this proves recently as the study of Vergel et al. [75] reported for the first time the presence of TSV in *P. vannamei* in the Philippines using morphological and molecular techniques. BLASTn search results showed that the TSV sequences have very high sequence similarity at 86–100% with TSV viral isolates from other countries (Taiwan, Thailand, Venezuela, the USA, Colombia, and Belize). The detected prevalence rates of the study comprise a small sample population with limited areas in the Philippines, namely, Bulacan (33%), Batangas (47%), Bohol (7%), and Cebu (13%). The authors suggested further testing in other sites in the country and implementation of mitigation methods and policies to prevent further spread of the viral disease. Likewise, detection of TSV in the wild is also important to be conducted. In the study of Barnette et al. [76] using PCR and immunological analyses, results suggested that TSV has already spread into the Bangpakong River and the Gulf of Thailand. The viral disease appeared to be more widespread in dry seasons than wet seasons. The presence of TSV has been detected in *P. monodon* adults; local shrimp species of the Bangpakong River such as *P. monodon*, *P. semisulcatus*, *P. merguiensis*, *M. brevicornis*, *M. affinis*, *M. tenuipes*, *Parapenopsis hungerfordi*, and *M. rosenbergii*; two other species belonging to the Family Caridea; and wild-caught *P. vannamei* including green mussel *Perna viridis*, blue swimming crab *Portunus pelagicus*, and Asian sea bass *Lates calcarifer*. TSV was detected in *L. setiferus* and *Farfantepenaeus aztecus* in Laguna Madre [77] and *L. schmitti* in Maracaibo lagoon, Venezuela [78].

## 7. Social impacts

With or without valid arguments, aquaculture has been accused to be the cause of many problems such as environmental, economic, inclusively esthetic, and social impacts [79]. In Vietnam, social impacts associated with shrimp farming include the increase of poverty and landlessness, food insecurity, and impacts on health and education [80, 81]. While there is no recent information regarding the social impacts of shrimp farming particularly the use of alien species such as the *P. vannamei* in the Philippines, Primavera [82] reported that the social costs of intensive prawn farming include the reduction of domestic and agricultural water supplies, decline in quantity of food fish, marginalization of coastal fishermen, displacement of labor, and credit monopoly by big businessmen. The capital-intensive nature of high-density shrimp culture has favored the entry of multinational corporate investors or the national elite. They can provide the necessary capital; have easier access to permits, credits, and subsidies; and can absorb financial risks which are disadvantaged to local communities in coastal areas and small farmers [83]. One main issue in the recent expansion of shrimp culture is social conflicts with other resource users; however, the shrimp culture industry employs thousands of rural people, and it would be far worse off without it [73].

## 8. Conclusion and recommendation

The presence of Pacific whiteleg shrimp *P. vannamei* in the Philippines has already reached more than five decades, and this raises ecological and social concern. This review paper aimed to evaluate the ecological and social impacts of aquacultural introduction to the Philippines waters of Pacific whiteleg shrimp *P. vannamei*. Several questions were answered and discussed based on literature, scientific details, reflections on personal experience, and their relevance to aquaculture of the *P. vannamei* in the Philippines to evaluate the ecological impacts, while social impacts were discussed based on literature. Findings revealed the escapes of *P. vannamei* from aquaculture production facilities of several countries into non-native waters including the Philippines. The presence of *P. vannamei* in the wild was due to the intentional release in attempts to escape detection during the implementation of a ban on the importation of all live shrimp in the Philippines and also possibly due to escapes from intensive and expanding production cycles as well as natural calamities such as floods. Consequently, the ability of *P. vannamei* escapees to survive the natural environmental conditions due to their tolerance to a wide range of salinity and pH could lead to ecological concerns such as resource competition, reproduction, and spread of disease in the wild. Studies reported that *P. vannamei* could potentially interact with local species through food competition, either by exploitative or interference competition. While there is no evidence that *P. vannamei* can establish population outside of its natural geographic range, natural reproduction of escapees is still possible once released mature adults could find mates in the wild. Thus, if the frequency of escapes is increasing and when they begin to reproduce successfully, the non-native shrimp *P. vannamei* could become a serious threat to native shrimps. The most problematic consequence of *P. vannamei* farming is the spread of the alien pathogen, Taura syndrome virus (TSV), which is rapidly spread due to international trade and now widely distributed in the shrimp-farming regions of the Americas, Southeast Asia, and the Middle East. The proliferation of TSV in *P. vannamei* farming has resulted in catastrophic losses and transmission of the disease in the wild wherein crustaceans (other shrimps and crabs) including fish (Asian sea bass) can be infected. In the Philippines, TSV presence in farmed *P. vannamei* was reported for the first time since its introduction in the 1970s. On the other hand, there is no specific information on the social impacts of *P. vannamei* farming; however, the recent expansion of shrimp culture has resulted in social conflicts with other resource users.

This review shows the negative implications on the aquacultural introduction of Pacific whiteleg shrimp *P. vannamei* in the Philippines both ecological and social aspects, and this heightens important management issues. Below are recommendations to ensure sustainable farming of *P. vannamei* in the Philippines:

1. Strict implementation of the guidelines for the importation and culture of the shrimp based on Fisheries Administrative Order No. 225, Series of 2007. Illegal *P. vannamei* farming must be prohibited.

2. Strengthen the screening requirements for the importation of all the shrimp stages, accreditation of hatchery, and grow-out facilities to eliminate the spread of TSV and escapees into the natural ecosystem. Monoculture of the shrimp must be one of the critical requirements for grow-out farm accreditation. Accredited hatcheries must only allow selling seedlings to accredited grow-out farm operators and for research purposes to eliminate illegal farming.

3. A collaborative program must be implemented on the nationwide detection and prevalence of TSV by concerned agencies such as DA-BFAR, DOST-PCAARRD, SEAFDEC/AQD, and SCU's in cooperation with private shrimp growers.

4. Establishment of a monitoring program for the presence of the shrimp and TSV in the wild to detect the geographic spread of escapees and the disease as well as their impacts.

5. Conduct social impacts of expanding *P. vannamei* farming.

6. Continuous support of the government for research to create sustainable aquaculture production of the shrimp.

## Author details

Marlon S. Alejos[1*], Augusto E. Serrano, Jr.[1,2], Yashier U. Jumah[1,3], Rey dela Calzada[1], Cyril Tom B. Ranara[1,4] and Jumari C. Fernandez[5]

1 Institute of Aquaculture, College of Fisheries and Ocean Sciences, University of the Philippines Visayas, Miagao, Iloilo, Philippines

2 National Institute of Molecular Biology and Biotechnology, University of the Philippines Visayas, Miagao, Iloilo, Philippines

3 Aquaculture Department, College of Fisheries, Mindanao State University Tawi-Tawi College of Technology and Oceanography, Sanga-Sanga, Bongao, Tawi-Tawi, Philippines

4 Institute of Aquatic and Applied Sciences, Davao Del Norte State College, Panabo City, Davao Del Norte, Philippines

5 Institute of Marine Fisheries and Oceanology, College of Fisheries and Ocean Sciences, University of the Philippines Visayas, Miagao, Iloilo, Philippines

*Address all correspondence to: msa0585@yahoo.com

## References

[1] Morton B. The aquatic nuisance species problem: A global perspective review. In: D'Itri FM, editor. Zebra Mussels and Aquatic Nuisance Species. Chelsea, Michigan: Ann Arbor Press; 1997. pp. 1-54

[2] International Union for Conservation of Nature (IUCN). Guidelines for the Prevention of Biodiversity Loss caused by Alien Invasive Species. 2000. Available from: https://www.iucn.org/

[3] Convention on Biological Diversity (CBD). About invasive alien species. Secretary-General of the United Nations. Available from: https://www.cbd.int/invasive/WhatareIAS.shtml

[4] Millennium Ecosystem Assessment. Ecosystems and Human Well-Being: Synthesis. Washington, DC: Island Press; 2005

[5] Invasive Special Specialist Group (ISSG). Review of the impact of invasive alien species on species protected under the convention on migratory species (CMS). In: For the Convention on the Conservation of Migratory Species of Wild Animals (CMS). Germany: United Nations Premises, Bonn; 2014

[6] India Environment Portal. Available from: www.indiaenvironmentportal.org.in/node/38152

[7] Bagarinao TU. The decline of native fishes and fisheries and the rise of aquaculture in lakes and Rivers in the Philippines. In: Santiago CB, Cuvin-Aralar ML, Basiao ZU, editors. In Conservation and Ecological Management of Philippine Lakes in Relation to Fisheries and Aquaculture. Southeast Asian Fisheries Development Center, Aquaculture Department, Iloilo, Philippines; Philippine Council for Aquatic and Marine Research and Development, Los Baños, Laguna, Philippines; and Bureau of Fisheries and Aquatic Resources: Quezon City, Philippines; 2001. p. 187

[8] Guerrero RD. Status of Alien/Invasive Fish Species in the Philippines. Paper presented at the Seminar-Workshop on Biodiversity and Management of Alien/Invasive Species in the Philippines, May 22-23, 2001, Quezon City. Protected Areas and Wildlife Bureau and ASEAN Regional Centre for Biodiversity Conservation; 2001

[9] Casal M. Documenting the presence and impact of introduced freshwater fishes for aquaculture in the Philippines. In: Cuvin-Aralar ML, Punongbayan RS, Santos-Borja A, Castillo LV, Manalili EV, Mendoza MM, editors. Proceedings of the First National Congress on Philippine Lakes (LakeCon 2003). Los Baños, Laguna, Philippines: SEAMEO SEARCA; 2005. p. 340

[10] FishBase, 2004. Entry for Aristichthys nobilis. In: Masuda H, Amaoka K, Araga C, Uyeno T, Yoshino T, editors. The Fishes of the Japanese Archipelago. Vol. 1. Tokyo, Japan: Tokai University Press; 1984. Available from: http://www.fishbase.org/

[11] Cagauan AG. Exotic aquatic species introduction in the Philippines for aquaculture – A threat to biodiversity or a boon to the economy? Journal of Environmental Science and Management. 2007;**10**(1):48-62

[12] Invasive Special Specialist Group (ISSG). Global Invasive Species Database. Invasive species. 2018. Available from: http://issg.org/database/species

[13] Ma C. Giant Tiger Prawn *Penaeus monodon*, Whiteleg Shrimp Litopenaeus Vannamei. Cannery Row Monterey, CA: Monterey Bay Aquarium's Seafood Watch; 2015. https://www.seafoodwatch.org/

[14] Boone. Penaeus vannamei. World register of marine species. 1931. Available from: http://www.marinespecies.org/aphia.php?p=taxdetails&id=377748

[15] Perez-Farfante I, Kensley B. Penaeoid and Sergestoid shrimps and prawns of the world, keys, and diagnoses for the species and genera. Memoires Museum National d' Histoire Naturalle, Zoologie, Tome. 1997;**175**:235

[16] Fortes RD. Aquaculture commodity imports: issues and problems. Paper presented at the Roundtable Dialogue on the Review of Fishery Policies, Department of Science and Technology, Taguig, Metro Manila, December 12-13; 1989

[17] Primavera JH. Aquaculture and the coastal environment. In: Paper Presented during the Seminar-Workshop on the Management of Nearshore Fisheries, 23-25 January 1989. Cebu City: Philippines; 1989

[18] Primavera JH. A critical review of shrimp pond culture in the Philippines. Reviews in Fisheries Science. 1993;**1**(2):151-201. DOI: 10.1080/10641269309388539

[19] Stern PC, Fineberg HV, editors. Understanding risk: Informing decisions in a democratic society. Choice Reviews Online. 1996, 1997;**34**(10):34-5653. DOI: 10.5860/choice.34-5653

[20] Hayes KH, Kapuscinski AR, Dan G, Li S, Devlin RH. Introduction to environmental risk assessment for transgenic fish. In: Kapuscinski AR, Hayes KH, Li S, Dan G, editors. Environmental Risk Assessment of Genetically Modified Organisms. Methodologies for Transgenic Fish. Vol. 3. Wallingford, UK: CAB International; 2007. pp. 1-28

[21] Senanan W, Panutrakul S, Barnette P, Manthachitra V, Chavanich S, Kapuscinski AR, et al. Ecological risk assessment of an alien aquatic species: A case study of Litopenaeus vannamei (pacific whiteleg shrimp) aquaculture in the Bangpakong River, Thailand. In: Tropical Deltas and Coastal Zones: Food Production, Communities and Environment at the Land-Water Interface. Oxfordshire, UK: CABI Publishing; 2010. pp. 64-79. Available from: https://www.cabi.org/what-we-do/cabi-centre/wallingford/

[22] Senanan W, Panutrakul S, Barnette P, Chavanich S, Manthachitra V, Tangkrock-Olan N, et al. Preliminary risk assessment of Pacific whiteleg shrimp (*P. vannamei*) introduced to Thailand for aquaculture. Aquaculture Asia Magazine. 2009;**14**:28-32

[23] Balboa WA, King TL, Hammerschmidt PC. Occurrence of Pacific white shrimp in Lower Laguna Madre, Texas. Proceedings of the Annual Conference Southeast Association Fish and Wildlife Agencies. 1991;**45**:288-292

[24] Brock JA. A brief overview of some historical and current issues concerning Penaeid shrimp introductions into Hawaii. In: DeVoe MR, editor. Introductions and Transfers of Marine Species. USA: South Carolina Sea Grant Consortium; 1992. pp. 187-189. Available from: https://www.scseagrant.org/contact-us/

[25] Eldridge LG. Introduction of commercially significant aquatic organisms to the Pacific Islands. Penaeid Shrimps pp. 91-95. SPC Inshore Fisheries Research Project Technical Document No. 7. SPREP Reports and Studies Series No. 78. ISSN: 1018-3116; 1995

[26] Howells R. Introduced non-native fishes and shellfishes in Texas waters: An updated list and discussion. Texas Parks

and Wildlife Department, Management Data Series 188; 2001. p. 27

[27] O'Connor M, Hawkins C, Loomis DK. A manual of previously recorded non-indigenous invasive and native transplanted animal species of the Laurentian Great Lakes and the coastal United States. NOAA Technical Memorandum NOS NCCOS 77. 2008. p. 82. Available from: http://aquaticcommons.org/2231/1/techmemo77.pdf

[28] Wenner EC, Knott DM. Occurrence of pacific white shrimp, Penaeus vannamei, in coastal waters of South Carolina. In: DeVoe MR, editor. Proceedings of the conferences and workshop: introductions and transfers of marine species: Achieving a balance between economic development and resource protection. Unpublished report, South Carolina Sea Grant Consortium; 1992. pp 173-181

[29] Briggs M, Funge-Smith S, Subasinghe R, Phillips M. Introductions and movement of *Penaeus vannamei* and *Penaeus stylirostris* in Asia and the Pacific. RAP Publication. 2004;**75**:1-12. DOI: 10.1016/j.wneu.2010.12.045

[30] Senanan W, Tangkrock-Olan N, Panutrakul S, Barnette P, Wongwiwatanawute C, Niphonkit N, et al. The presence of the pacific whiteleg shrimp (*Litopenaeus vannamei*, Boone, 1931) in the wild in Thailand. Journal of Shellfish Research. 2007;**26**(4):1187-1192

[31] Pérez JE, Alfonsi C, Salazar SK, Macsotay O, Barrios J, Martínez Escarbassiere R. Especies marinas exóticas y criptogénicas en las Costas de Venezuela. Boletin del Instituto de Oceanografia Venezuela. 2007;**46**(1):79-96

[32] Loebmann D, Mai ACG, Lee JT. The invasion of five alien species in the Delta do Parnaíba environmental protection area, Northeastern Brazil. Revista de Biología Tropical. 2010;**58**(3):909-923. DOI: 10.15517/rbt.v58i2.5254

[33] Perry H. *Litopenaeus vannamei*. USGS Non-indigenous Aquatic Species Database. Gainsville, Florida: United States Geological Survey; 2011

[34] Binh LT, Yen MD, Luyen NH. Preliminary impacts assessment of alien aquatic species on biodiversity as well as invasion of native fishes in aquaculture and some management measures. In: 13th World Lake Conference. Rehabilitate the Lake Ecosystem: Global Challenges and the Chinese Innovations; 2009

[35] Wakida-Kusunoki AT, Amador-del Angel LE, Alejandro PC, Brahms CQ. Presence of pacific white shrimp *Litopenaeus vannamei* (Boone, 1931) in the southern Gulf of Mexico. Aquatic Invasions. 2011;**6**(1):S139-S142. DOI: 10.3391/ai.2011.6.S1.031

[36] Rosario WR, Lopez NA. Status of P. vannamei aquaculture in the Philippines. In: Regional Technical Consultation on the Aquaculture of *P. vannamei* and Other Exotic Shrimps in Southeast Asia, Manila, Philippines. Tigbauan, Iloilo, Philippines: SEAFDEC Aquaculture Department; 2005. pp. 62-68. Available from: http://hdl.handle.net/10862/853

[37] Cuvin-Aralar MLA, Lazartigue AG, Aralar EV. Cage culture of the Pacific white shrimp *Litopenaeus vannamei* (Boone, 1931) at different stocking densities in a shallow eutrophic lake. Aquaculture Research. 2009;**40**(2):181-187. DOI: 10.1111/j.1365-2109.2008.02081.x

[38] Philippine Shrimp Congress. Philippine Shrimp Congress (6th) slated in Bacolod City. Visayan Daily Headlines; 2008. pp. 3-6

[39] BFAR. Master List of Shrimp Farms as of December 2013; 2013

[40] BFAR. List of Accredited Hatcheries of P. Vanamei of as 2014; 2014

[41] DA-BFAR. List of BFAR Accredited/Certified Specific Pathogen Free (SPF) Penaeus vannamei Grow-out Farms as of 31 August 2019; 2019a

[42] DA-BFAR. List of Registered Penaeus Vannamei Extensive Grow-out Farms as of 31 August 2019; 2019c

[43] DA-BFAR. List of Accredited SPF Penaeus vannamei Hatcheries as of 31 August 2019; 2019b

[44] Panutrakul S, Senanan W, Chavanich S, Tangkrock-Olan N, Viyakarn V. Ability of *Litopenaeus vannamei* to survive and compete with local marine shrimp species in the Bangpakong River, Thailand. In: Tropical Deltas and Coastal Zones: Food Production, Communities, and Environment at the Land-Water Interface. Oxfordshire, UK: CABI Publishing; 2010. pp. 80-92. Available from: https://www.cabi.org/what-we-do/cabi-centre/wallingford/

[45] Manthachitra V, Phadejpai K, Intachreon P, Jeranaiwiwat S. An assessment of land used changes of Pacific whiteleg shrimp (*Litopenaeus vannamei*) culture in the Bangpakong watershed using geographic information system. In: Panutrakul S, Senanan W, editors. Aquaculture Management Strategies for the Pacific Whiteleg Shrimp (Litopenaeus vannamei) in the Bangpakong River Basin and the East Coast of Thailand. Chon Buri: Burapha University; 2008. pp. 11-35

[46] Duan Y, Zhang Y, Dong H, Zheng X, Wang Y, Li H, et al. Effect of dietary poly-β-hydroxybutyrate (PHB) on growth performance, intestinal health status and body composition of Pacific white shrimp Litopenaeus vannamei (Boone, 1931). Fish & Shellfish Immunology. 2017;**60**:520-528. DOI: 10.1016/j.fsi.2016.11.020

[47] Yang SP, Wu ZH, Jian JC, Zhang XZ. Effect of marine red yeast Rhodosporidium paludigenum on growth and antioxidant competence of Litopenaeus vannamei. Aquaculture. 2010;**309**(1-4):62-65. DOI: 10.1016/j.aquaculture.2010.09.032. Available from: http://bagong.pagasa.dost.gov.ph/

[48] Texas Parks and Wildlife Department. Exotic shrimp confirmed in Matagorda Bay. News Release; 1997

[49] Wenner EL, Knott DM. Occurrence of Pacific white shrimp, Penaeus vannamei, in coastal waters of South Carolina. In: Proceedings of the Conference and Workshop, Introduction and Transfer of Marine Species. Achieving a Balance between Economics and Development and Resources Protection. South Carolina; 1991. pp. 173-181

[50] Medina-Reyna CE. Growth and emigration of white shrimp, *Litopenaeus vannamei*, in the mar Muerto lagoon, southern Mexico. Naga. 2001;**24**(3/4):30-34

[51] Chavanich S, Voranop V, Senanan W, Panutrakul S. Ecological impacts of Pacific whiteleg shrimp (*Litopenaeus vannamei*) aquaculture on native shrimps and crabs in Bangpakong watershed. In: Panutrakul S, Senanan W, editors. Aquaculture Management Strategies for the Pacific Whiteleg Shrimp (Litopenaeus vannamei) in the Bangpakong River Basin and the East Coast of Thailand. Chon Buri: Burapha University; 2008. pp. 139-153. (in Thai with English abstract)

[52] CABI. Litopenaeus vannamei Boone 1931. In: Invasive Species Compendium. Wallingford, UK: CAB International; 2019. Available from: www.cabi.org/isc

[53] Dugassa H, Gaetan DG. Biology of white leg shrimp, Penaeus vannamei: Review. World Journal of Fish and

Marine Sciences. 2018;**10**(2):5-17. DOI: 10.5829/idosi.wjfms.2018.05.17

[54] Bailey-Brock JH, Moss SM. Penaeid taxonomy, biology, and zoogeography. Marine Shrimp Culture. Elsevier; 1992;**23**:9-27. DOI: 10.1016/b978-0-444-88606-4.50008-4 Available from: https://www.elsevier.com/books/marine-shrimp culture/fast/978-0-444-88606-4

[55] Senanan W, Barnette P, Ponsombut K, Tangkrock-Olan N. Gonadal development of Pacific whiteleg shrimp (*Litopenaeus vannamei*) raised in captivity and caught from the wild. In: Panutrakul S, Senanan W, editors. Aquaculture Management Strategies for the Pacific Whiteleg Shrimp (*Litopenaeus vannamei*) in Bangpakong River Basin and the East Coast of Thailand. Chon Buri: Burapha University; 2008. pp. 51-57

[56] Chavanich S, Viyakarn V, Senanan W, Panutrakul S. Laboratory assessment of feeding-behavior interactions between the introduced pacific white shrimp *Litopenaeus vannamei* (Boone, 1931) (Penaeidae) and five native shrimps plus a crab species in Thailand. Aquatic Invasions. 2016;**11**(1):67-74. DOI: 10.3391/ai.2016.11.1.07

[57] Gamboa-Delgado J, Molina-Poveda C, Cahu C. Digestive enzyme activity and food ingesta in juvenile shrimp *Litopenaeus vannamei* (Boone, 1931) as a function of body weight. Aquaculture Research. 2003;**34**(15):1403-1411. DOI: 10.1111/j.1365-2109.2003.00959.x

[58] Ruiz GM, Fofonoff PW, Carlton JT, Wonham MJ, Hines AH. Invasion of coastal marine communities in North America: Apparent patterns, processes, and biases. Annual Review of Ecology and Systematics. 2000;**31**:481-531 http://dx.doi.org/10.1146/annurev.ecolsys.31.1.481

[59] Williamson M. Biological Invasions. U. K.: Chapman and Hall; 1996. p. 244

[60] Anderson JL, Valderrama D, Jory D. Shrimp Production Review. United States: Global Aquaculture Alliance; 2017. Available from: https://www.bloomberg.com/profile/company/1401676D:US

[61] Andrino-Felarca KGS, Estante EG, Lazado CC. 2. Viral diseases of shrimp in the Philippines. In: Caipang CMA, Bacano-Maningas MBI, Fagutao FF, editors. Biotechnological Advances in Shrimp Health Management in the Philippines. Vol. 661(2). Trivandrum, Kerala, India; 2015. pp. 19-44. ISBN: 978-81-308-0558-0

[62] Itsathitphaisarn O, Thitamadee S, Weerachatyanukul W, Sritunyalucksana K. Potential of RNAi applications to control viral diseases of farmed shrimp. Journal of Invertebrate Pathology. 2017;**147**:76-85. DOI: 10.1016/j.jip.2016.11.006

[63] Lightner DV, Redman RM. Strategies for the control of viral diseases of shrimp in the Americas. Fish Pathology. 1998;**33**(4):165-180. DOI: 10.3147/jsfp.33.165

[64] Lightner DV. Taura syndrome: An economically important viral disease impacting the shrimp farming industries of the Americas including the United States. Proceedings, Annual Meeting of the United States Animal Health Association. 1995;**99**:36-52

[65] Lightner DV. The Penaeid shrimp viruses TSV, IHHNV, WSSV, and YHV: Current status in the Americas, available diagnostic methods, and management strategies. Journal of Applied Aquaculture. 1999;**9**(2):27-52. DOI: 10.1300/J028v09n02_03

[66] Lightner DV, editor. A Handbook of Shrimp Pathology and Diagnostic Procedures for Diseases of Cultured

Penaeid Shrimp. Baton Rouge, Louisiana, USA: World Aquaculture Society; 1996. 304 pp

[67] Wertheim JO, Tang KFJ, Navarro SA, Lightner DV, editor. A quick fuse and the emergence of Taura syndrome virus. Virology. 2009;**390**(2):324-329. DOI: 10.1016/j.virol.2009.05.010

[68] Lightner DV. Taura syndrome. Manual of Diagnostic Tests for Aquatic Animals. 2006;**33**:16-378. Retrieved from: http://search.proquest.com/docview/920802945?accountid=14643

[69] Walker PJ, Winton JR. Emerging viral diseases of fish and shrimp. Veterinary Research. 2010;**41**(6):51. DOI: 10.1051/vetres/2010022

[70] Flegel TW. Current status of viral diseases in Asian shrimp aquaculture. The Israeli Journal of Aquaculture-Bamidgeh. 2009;**61**(3):229-239

[71] Thitamadee S, Prachumwat A, Srisala J, Jaroenlak P, Salachan PV, Sritunyalucksana K, et al. Review of current disease threats for cultivated penaeid shrimp in Asia. Aquaculture. 2016;**452**:69-87. DOI: 10.1016/j.aquaculture.2015.10.028

[72] NACA, OIE, FAO. Network of Aquaculture Centers in Asia-Pacific (NACA), World Organization for Animal Health (OIE) Regional Representation for Asia and the Pacific, and Food and Agriculture Organization of the United Nations (FAO). Quarterly Aquatic Animal Disease Report (Asia and Pacific Region), 2018/4, October–December 2018. NACA, Bangkok, Thailand and OIE-RRAP, Tokyo, Japan; 2019

[73] FAO. Cultured aquatic species information programme. *Penaeus vannamei*. Cultured aquatic species information programme. Text by Briggs M. In: FAO Fisheries and Aquaculture Department [online]. 2006-2019. Rome. Updated: 7 April 2006. Available from: http://www.fao.org/fishery/culturedspecies/Penaeus_vannamei/en [Cited: 20 October 2019]

[74] de la Peña LD. Transboundary shrimp viral diseases with emphasis on white spot syndrome virus (WSSV) and Taura syndrome virus (TSV). In: Lavilla-Pitogo CR, Nagasawa K, editors. Transboundary Fish Diseases in Southeast Asia: Occurrence, Surveillance, Research, and Training. Proceedings of the Meeting on Current Status of Transboundary Fish Diseases in Southeast Asia: Occurrence, Surveillance, Research and Training, Manila, Philippines, 23-24 June 2004. Tigbauan, Iloilo, Philippines: SEAFDEC Aquaculture Department; 2004. pp. 67-69

[75] Vergel JV, Cabawatan LP, Madrona VC, Rosario AT, Sta. Ana JBM, Tare MR, et al. Detection of Taura syndrome virus (TSV) in *Litopenaeus vannamei* in the Philippines. The Philippine Journal of Fisheries. 2019;**26**(1):8-14. DOI: 10.31398/tpjf/25.2.2018-0004

[76] Barnette P, Senanan W, Khururat O, Wongwiwatanawute C. Distribution and transmission of virus diseases between (Litopenaeus vannamei) and native shrimp species in Bangpakong watershed. In: Panutrakul S, Senanan W, editors. Aquaculture Management Strategies for the Pacific Whiteleg Shrimp (Litopenaeus vannamei) in the Bangpakong River Basin and the East Coast of Thailand. Chon Buri: Burapha University; 2008. pp. 106-138. (in Thai with English abstract)

[77] Guzmán-Sáenz FM, Pérez-Castañeda R, Molina-Garza ZJ, Ibarra-Gámez JC, Galavíz-Silva L. Infectious hypodermal and hematopoietic necrosis virus (IHHNV) and Taura syndrome virus (TSV)

in wild shrimp (*Farfantepenaeus aztecus* Ives, 1891 and *Litopenaeus setiferus* Linnaeus, 1767) of La Laguna Madre, Gulf of Mexico. Revista de Biología Marina y Oceanografía. 2009;**44**(3):663-672. DOI: 10.4067/S0718-19572009000300012

[78] Fajardo C, Rodulfo H, De Donato M, Manrique R, Boada M, Aguado N. Detección molecular del virus del síndrome del Taura en litopenaeus schmitti silvestres del lago de Maracaibo y la laguna de Unare, Venezuela. Revista Cientifica de La Facultad de Ciencias Veterinarias de La Universidad Del Zulia, 2010;**20**(5):457-466

[79] Martinez-Porchas M, Martinez-Cordova LR. World aquaculture: Environmental impacts and troubleshooting alternatives. The Scientific World Journal. 2012;**2012**:389623. DOI: 10.1100/2012/389623

[80] Primavera JH, Barraclough S, Finger-Stich A. Some Ecological and Social Implications of Commercial Shrimp Farming in Asia. Ocean and Coastal Management. 2006;**49**(66):9-10. DOI: 10.1046/j.1365-2109.1997.00946.x

[81] Preston N, Clayton H. Rice-shrimp farming in the Mekong Delta: Biological and Socioeconomic Issues, ACIAR (Australian Center for International Agricultural Research) Technical reports no. 52e; 2003

[82] Primavera JH. Intensive prawn farming in the Philippines - Ecological, social, and economic-implications. Ambio. 1991;**20**(1):28-33

[83] Primavera JH. Socio-economic impacts of shrimp culture. Aquaculture Research. 1997;**28**(10):815-827. DOI: 10.1111/j.1365-2109.1997.tb01006.x

Section 4

# Landscape Sciences

Chapter 8

# High-Resolution Object-Based Building Extraction Using PCA of LiDAR nDSM and Aerial Photos

*Alfred Cal*


## Abstract

Accurate and precise building extraction has become an essential requirement for various applications such as for impact analysis of flooding. This chapter seeks to improve the current and past methods of building extraction by using the principal components analysis (PCA) of LiDAR height (nDSM) and aerial photos (in four RGB and NIR bands) in an object-based image classification (OBIA). This approach uses a combination of aerial photos at 0.1-m spatial resolution and LiDAR nDSM at 1-m spatial resolution for precise and high-resolution building extraction. Because aerial photos provide four bands in the PCA process, this potentially means that the resolution of the image is maintained and therefore building outlines can be extracted at a high resolution of 0.1 m. A total of five experiments was conducted using a combination of different LiDAR derivatives and aerial photos bands in a PCA. The PCA of LiDAR nDSM and RGB and NIR bands combination has proved to produce the best result. The results show a completeness of 87.644%, and a correctness of 93.220% of building extraction. This chapter provides an improvement on the drawbacks of building extraction such as the extraction of small buildings and the smoothing with a well-defined building outline.


**Keywords:** building foot prints, LiDAR, nDSM, principal components analysis, object-based image classification

## 1. Introduction

Over the years, buildings extraction at a high resolution has become an essential requirement for various applications such as flood modeling, urban planning, and 3D building modeling. In flood modeling scenarios, one of the most important structures at risk are buildings. Buildings houses people and other valuable assets, therefore, proper representation of buildings is very important for flood managers. Currently, building extraction methods are being done using a mixture of different data sources and various algorithms. The use of high-resolution aerial imagery and LiDAR are commonly integrated for more accurate building extraction. Aerial image provides spectral information, while LiDAR data provides height and intensity information. By fusing 2D aerial images and 3D information from LiDAR, complementary information can be exploited to improve automatic building extraction processing and the accuracy of the building roof outline [1].

There are several techniques used for building extraction in the remote sensing field. One such technique is called image fusion. Image fusion is the combination of two or more different images to form a new image by using an algorithm to obtain more and better information about an object or a study area [2]. Multispectral data such as aerial photos has spectral and high spatial resolution, meanwhile, LiDAR data has height and intensity information. Thus, buildings can be extracted based on their height from LiDAR and spectral information from aerial photos to improve the spatial resolution of roof edges [3].

There are many image fusion methods that are available, these include intensity hue-saturation (IHS), Brovey transform (BV) and principal component analysis (PCA) [2]. PCA transformation is a technique to reduce the dimensionality of multivariate data whilst preserving as much of the relevant information as possible. It also translates correlated data set to uncorrelated dataset [2]. In this study, a fusion of aerial photos and LiDAR datasets using PCA can be beneficial to accurately detect and extract buildings at a high spatial resolution. The advantage of using PCA as an image fusion technique for feature extraction is that all resulting variables are independent of each other while still retaining the most valuable parts of the input variables. Thus, other type of transformations, such as IHS destroys the spectral characteristics of the image data which is important for feature extraction and Brovey Transform depress the image values during image fusion [2].

For building extraction, some form of image classification technique is needed. Pixel-based (spectral pattern recognition), and object-based (spatial pattern recognition) are the two groups of common image classification techniques. Traditionally, image classifications are done with pixel-base using different classifiers in supervised and unsupervised classification (e.g., K-Means, Maximum Likelihood, etc.). These pixel-based procedures analyze the spectral properties of every pixel within the area of interest, without taking into account the spatial or contextual information related to the pixel of interest [4]. Meanwhile, object-based classification techniques start by grouping of neighboring pixels into meaningful areas. Object-based feature extraction is a relatively modern technique for the extraction of objects in urban environments such as buildings and roads, where its advantage lies in the classification of objects represented by a group of pixels. More specifically, image objects are groups of pixels that are similar to one another based on a measure of spectral properties (i.e., color), size, shape, and texture, as well as context from a neighborhood surrounding the pixels [5].

The goal of this chapter is to improve on past and existing methods of building extraction by introducing the use of principal component analysis (PCA) of LiDAR height (nDSM) and aerial photos (RGB and NIR) in an OBIA. This approach was evaluated by comparing the accuracy and quality of building extraction on 5 PCA datasets, this incudes (1) PCA combinations of RGB and nDSM, (2) PCA of RGB, NIR, nDSM and slope, (3) PCA of RGB, nDSM and NDVI, (4) PCA of RGB, NIR, nDSM and NDVI, and (5) PCA of RGB, NIR and nDSM.

By evaluating the combinations of bands using the PCA approach for building extraction, the author seeks to answer the research question of this study. To investigate which PCA parameter has the most influence for the detection and extraction of buildings and to determine which band combinations can produce a satisfactory building extraction in terms of their completeness, correctness, and quality. This approach provides the extraction of buildings at a high spatial resolution of 1-m, this allow the building outline to be extracted at the same high spatial resolution. This study introduces a novel approach of using PCA for precise and high-resolution building footprints extraction in an object-based image classification technique in a semi-automated process. Furthermore, this approach was validated by comparing the resultant building footprints using a quantitative and qualitative statistical analysis discussed in the results section.

The proposed method has been tested on a 1 $km^2$ area of Vista del Mar, Ladyville village, Belize Central America. The area chosen has a relatively flat landscape that has a combination of different building sizes and shapes, vegetation cover and waterbodies. The datasets used in this study are LiDAR point cloud and aerial photos. The LiDAR datasets and aerial photos for this study were made possible from the Ministry of Works (MOW), Belize and described in more detail in Chapter 3.

The rest of the chapter is organized as follows: Section 2 presents a background of building extraction from LiDAR data and multispectral images. Section 3 details the data and methods used for extraction. The results are discussed in Section 4. Finally, concluding remarks are offered in Section 5.

## 2. Background and related works

There have been many studies in building extraction techniques with different approaches. Some studies use only LiDAR data, others use only multispectral images and then there are those that use a combination of LiDAR and multispectral images to extract building outlines for various applications. Several methods have been presented for building extraction from LIDAR data during the last decades. Based on the used data, building extraction methods can generally be divided into three categories: 2D (two-dimensional) imagery based, fused 2D-3D information based, and 3D LiDAR based [1].

Studies that use only LiDAR data for building extraction includes [1, 6–9]. A typical step of combining geometry features to extract building is, firstly to filter the DTM from LIDAR data, then derive the DSM data into ground points and non-ground points (including vegetation and building) by height difference [7]. DSM is normally used in flood modeling applications with the combination of DEM to derive a difference image, also called normalize height image. The difference image is a result of subtracting the DEM from the DSM to get the absolute height of buildings and trees in the study area. Digital Surface Models (DSMs) offer the possibility of extracting the elevations of surface features to leave the ground surface DEM [9]. Airborne LiDAR Laser Scanning devices can provide digital surface models that can be used to separate surface features from the ground for modeling flood inundation from rivers in urban and semi-urban environments [9]. However, 3D information provided by LiDAR cannot solve all automated building extraction problems. A typical example is that to separate nearby trees and buildings, extra information, such as color or brightness, is needed to separate features [8]. In the extraction process, only those buildings that are classified as buildings in the point cloud data are extracted, in some cases tiny or small buildings that are not classified in the point cloud data will not be extracted. Another drawback is that this method still has some deficiency to extract out some very small building information. Further improving and updating is still necessary [7]. However, since the method uses LIDAR data alone, the planimetric accuracy is limited by the LIDAR point density. At present, the method does not incorporate smoothing of the boundaries of extracted planar segments [6]. And it is hard to obtain a detailed and geometrically precise boundary using only LIDAR point clouds [10].

Studies that used only images for building extraction includes [3, 11–13]. In remote sensing, building extraction from high resolution imagery has been a common field of research. So far, many algorithms have been presented for the extraction of buildings from satellite images and aerial photos. These algorithms have mainly considered radiometric, geometric, edge detection and shadow criteria approaches [13]. Although promising results have been obtained from these 2D information-based methods, shadows and occlusions leading to significant errors, especially in densely developed areas, cannot be avoided. Consequently, these methods are considered to be insufficiently automated and reliable for practical applications [6].

The third category of building extraction is using a combination of LiDAR data with multispectral images in an image fusion technique. This third approach exploits the mutual benefits of both datasets for accurate building extraction. By fusing 2D images and 3D information from LiDAR, complementary information can be exploited to improve automatic building extraction processing and the accuracy of the building roof outline [1]. This method has been widely studied in [14–16]; Building detection techniques integrating LIDAR data and imagery can be divided into two groups. Firstly, there are techniques that use the LIDAR data as the primary cue for building detection and those which use both the LIDAR data and the imagery as the primary cues to delineate building outlines [15]. In this approach LiDAR height and intensity are usually used along with aerial imagery to improve the classification of buildings. However, the challenges are how to integrate the two data sources for building boundary extraction still arises; few approaches with technical details have thus far been published [14].

## 3. Data and methods

The study area for this chapter is Ladyville Village, Belize, Central America. Ladyville was once a small coastal settlement separated from other communities, but over the years it has seen an increase in development and in population. Development has caused the village to become a sizable town and is sometimes considered a suburb of Belize City. Belize City, the largest city in Belize is only a few minutes' drive away from Ladyville with the Belize River separating the two settlements. Ladyville is north of Belize City, along the Belize River, along the coast, and along the Philip Goldson Highway and it is in the lower reach of the Belize River watershed. The study area map is provided in **Figure 1**.

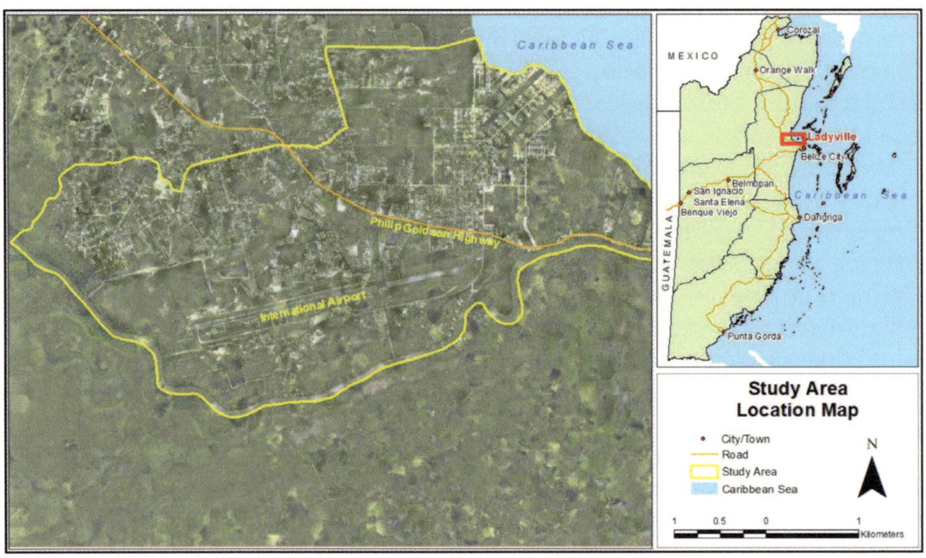


**Figure 1.**
*Study area location map.*

The topography of Ladyville is mostly flat. It is part of Belize's coastal lowland and it is a part of the Belize River natural floodplain. Its natural vegetation is mostly broadleaf lowland forests and marshlands with meandering creeks, lagoons, and mangrove forest along the coast. The Ladyville area was also a location where excavations were done to gather fill for sites in Belize City. Ladyville was chosen as a study area because it is one of the most vulnerable communities to natural disasters and because of its strategic importance. Ladyville is located between the Belize River and the Caribbean Sea. This means it is highly vulnerable from both river flooding from the Belize River and storm surge flooding from the Caribbean Sea from hurricanes or oceanic events.

The datasets used in this study are LiDAR and aerial photos. LiDAR point cloud data was provided in LAS format version 1.4. LAS is a standard data exchange format for LiDAR point cloud data established by (ASPRS) the American Society for Photogrammetry and Remote Sensing. The data has a point average spacing of 0.3 m and it was classified into ground, low vegetation, medium vegetation, high vegetation, buildings and noise. Aerial photos were taken within the same time period of LiDAR airborne surveys in August of 2017. The image has a high spatial resolution of 0.1 m (10 cm). Aerial photos have four bands, Red, Green, Blue and NIR (Near Infrared).

To complete the semi-automated building extraction process, the workflow was developed. The workflow shown in **Figure 2** below includes the following steps: *LiDAR Pre-processing, PCA, OBIA, Segmentation, Feature Extraction, Training Sites, Image Classification, Rule base Classification, Accuracy Assessment and Regularize Building Outline.*

### 3.1 LiDAR nDSM pre-processing

LiDAR pre-processing involves the filtering of ground points from non-ground points. As a result, two files with the digital elevation model (DEM) ground points only and the digital surface model (DSM) non-grounds points were created with a spatial resolution of 1 m. From the DEM and DSM, an nDSM was created by subtracting the DEM from the DSM. The normalized digital surface model (nDSM) represents the absolute height of objects in the study area such as buildings and trees. Then LiDAR nDSM as a separate band was combined with four aerial photos bands. LiDAR height information from the nDSM was added to the aerial photos which is an essential building characteristic for extraction from other features.

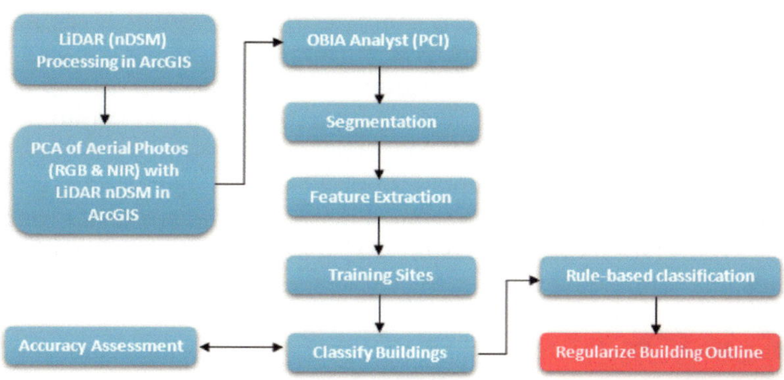


**Figure 2.**
*OBIA building extraction workflow diagram.*

### 3.2 PCA

Principal component analysis is a technique used to reduce the dimensionality of multivariate and multispectral datasets such as images with the aim of preserving as much of the relevant information as possible. PCA provides a method for the reduction of redundant information apparent in multi-dimensional databases. PCA represents any object with a much fewer information compared to the original image. Minimization of the correlation of multidimensional bands is performed by mathematically transforming the multi-band into another vector space with a new basis [17]. PCA was performed on the aerial photos in combination with LiDAR nDSM raster. The result is a single multiband raster, this means that the result of the LiDAR nDSM and aerial photos is a raster with 5 bands in a single raster dataset [18].

### 3.3 Object-based image classification (OBIA)

Object-based image classification (OBIA) is seen as an advancement in land cover classification, where its advantage lies in the classification of objects represented by a group of pixels. OBIA approaches for analyzing remotely sensed data have been established and investigated since the 1970s. Object-oriented methods of image classification have become more popular in recent years due to the availability of software [19]. Object-based classification techniques start by the grouping of neighboring pixels into meaningful areas. This means that the segmentation and subsequent object topology generation is controlled by the resolution and the scale of the expected objects. In an object-based classified image, the elementary picture elements are no longer the pixels, but connected sets of pixels [20].

### 3.4 Segmentation

The segmentation process in OBIA is used to recognize, differentiate and separate features within the image. This method involves the grouping of pixels into regions or areas based on their similar spectral reflectance, texture and area. Segmentation is defined as the delineation of the entire digital image into a number of segments or set of pixels, the goal is to enhance the present objects of the image into something more meaningful and required [21]. The segmentation process is dependent on the scale, shape, and compactness of objects. Several tests are needed to determine the best scale to use for image segmentation.

### 3.5 Feature extraction

The feature extraction process is performed after the image is segmented, this involves the searching of meaningful objects within the image such as roads, vegetation and buildings. This process allows us to isolate and extract only the object features that we need or that we are interested in. The computation of feature extraction can be statistical such as mean height, geometrical such as shape, elongation, rectangularity, and compactness. These parameters play an important role in the final output of extraction. The spatial and spectral properties are the two important factors for extraction [21]. The features extracted from the image bands or channels are used in the supervised classification of buildings.

### 3.6 Training sites

The training site section involves the selection of training sites for the building classification, the building features that are selected are those that have different characteristics such as color and shape. During the training site selection, other buildings can be selected that will be used for the accuracy assessment of the OBIA process. Buildings selected in the training site selection cannot be selected again for the accuracy assessment process.

### 3.7 Classification

Classification involves a supervised classification of the buildings for example using the support of vector machine (SVM). SVM has recently been given much attention as a classification method. In recent studies, Support vector machines were compared to other classification methods, such as Neural Networks, Nearest Neighbor, Maximum Likelihood and Decision Tree classifiers for remote sensing imagery and have surpassed all of them in robustness and accuracy [22].

### 3.8 Accuracy assessment

After classification, accuracy assessment is needed to determine the reliability of the classification process. This can be done by creating an accuracy assessment report or visually inspecting the results of the classification using the original image of the study area.

### 3.9 Rule-based classification

There is no classification or extraction process that is 100% accurate, therefore improvements can be made using rule-based classification. This involves making improvements to the results of the extraction process by using the attributes of the segmented layer. Geometrical rule-based classification involves selecting the desirable shape, compactness, rectangularity and elongation of objects, meanwhile statistical rule-based classification, involves selecting the mean height or mean NIR values from the segmented layer to improve the extraction of buildings.

### 3.10 Regularize building outlines

After the building extraction, the building outlines are observed to be very definitive at a large scale of 1:1000, which is significantly sufficient for various applications and scenarios. Nonetheless, zooming in closer at a very large scale of 1:250, some jagged edges can be seen. These minor rough or jagged edges were eliminated by cleaning the edges of buildings by choosing a standard precision and tolerance value to regularize the building outlines.

## 4. Results

Several experiments have been completed to determine the best combination of PCA raster data for the building extraction process. A total of five experiments have been completed to determine the best scenario of building/roof extraction on a 1 sq. km area. The aerial photograph of these areas shows a total of 584 buildings; therefore, the accuracy of building extraction was measured using this number. A total of 20 buildings are chosen for the training sites, these buildings sites are used in all five approaches. **Table 1** is provided further below that gives a quantitative analysis of the process.

For the building segmentation process (**Figure 3**), a scale of 25, shape 0.5 and Compactness 0.5 was used, this parameter creates much smaller segments for smaller objects such as buildings. The objects (polygon) layer created by the segmentation is accompanied by an attribute table containing a unique identification field for every object. Segmentation is completed only on the first three bands of the PCA raster, the first three bands is the equivalent of the RBG in aerial photos. **Figure 4** illustrates the size of the polygons used in the segmentation process, the building segments shown are smaller, in some cases 7 segments represents a building, this allows for a better building extraction with a well-defined building outline.

The first test was conducted using the PCA of RGB and nDSM. This PCA raster has a total of four bands, red, green, blue and the height data from the nDSM. Segmentation was completed on the RGB bands only, however, feature extraction is completed on all bands. Image segmentation is recommended only on the RGB bands which provide, the color, shape, and textures of objects in the study area. After feature attraction, all segments are given attribute information from all the four bands, these include the mean values of RGB and height data form the nDSM. Using this approach, it is observed that most buildings were selected, however, there are many other features that are selected as buildings, these features are those that have similar height of buildings such as vegetation and fences. In addition,

| PCA dataset | Commission percentage | Omission percentage | Completeness | Correctness | Quality |
|---|---|---|---|---|---|
| RGB and nDSM | 55.279% | 4.229% | 42.829% | 95.770% | 42.033% |
| RGB, NIR, nDSM and Slope | 40.336% | 42.055% | 59.663% | 57.944% | 41.634% |
| RGB, nDSM and NDVI | 43.828% | 2.417% | 56.171% | 97.582% | 54.650% |
| RGB, NIR, nDSM and NDVI | 37.608% | 3.563% | 62.391% | 96.436% | 60.985% |
| RGB, NIR and nDSM | 14.097% | 7.270% | 87.644% | 93.220% | 82.392% |

**Table 1.**
*Result of area-based accuracy measures.*

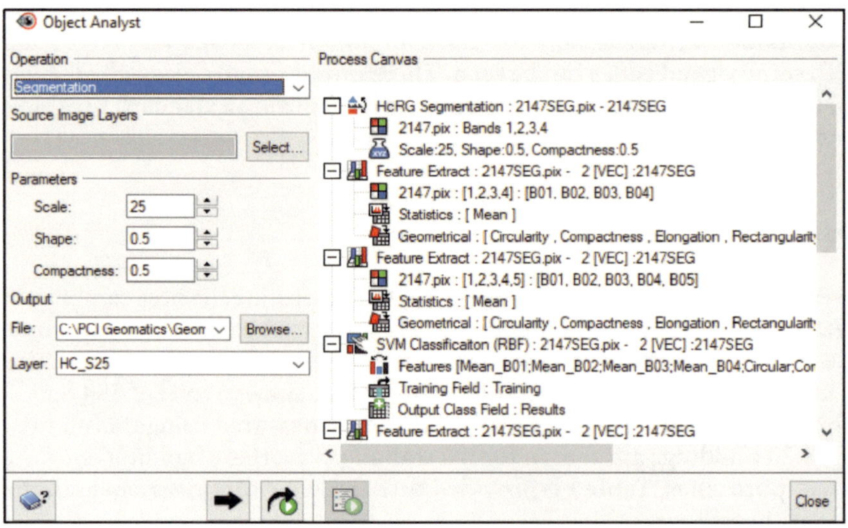


**Figure 3.**
*Segmentation parameters of the OBIA.*

**Figure 4.**
*The size and scale of the segments used in the building extraction.*

the outline of buildings is not well defined, although buildings are correctly classified, however, their shapes are not realistic of building outlines, this would require several editing and adjustments (**Figure 5**).

The second test involves the PCA of RGB, NIR, nDSM and slope, this raster data contains six bands. Slope is considered and additional parameter that can aid in roof extraction, however the result of this approach is poor as many buildings are not classified and those that are selected, their outlines were not smooth and definitive. It is observed that additional bands in the PCA slightly lowers the spatial resolution of the datasets and therefore objects are not well defined (**Figure 6**).

The third approach includes the PCA of RGB, nDSM and NDVI, a total of five bands. The NDVI (normalize difference vegetation index) is used in remote sensing to analyze the health of vegetation from green being healthy to red not healthy. NDVI was included to try to separate the objects that are green which are vegetation from other features that are not green such as buildings. The results (**Figure 7**) look promising where all buildings are selected, however, other features such as waterbodies and roads are classified as buildings, this observed to be because of the similar NDVI values of roads and waterbodies with the buildings. A closer observation shows that buildings that are close to each other are selected as one building and most of their outline is not well defined.

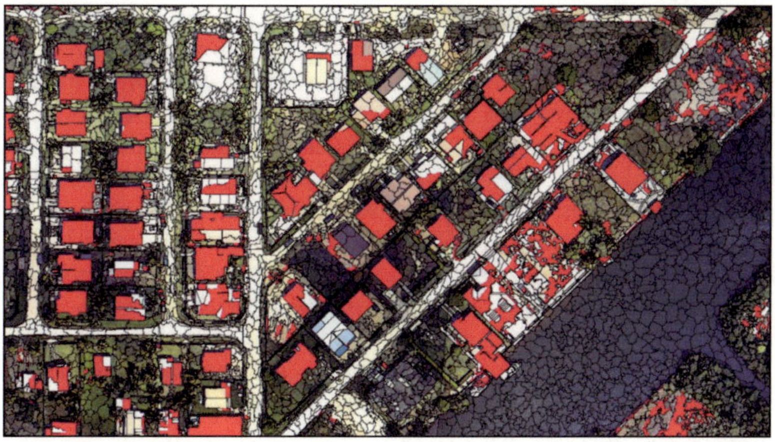

**Figure 5.**
*Building footprint extraction using PCA of RGB and nDSM.*

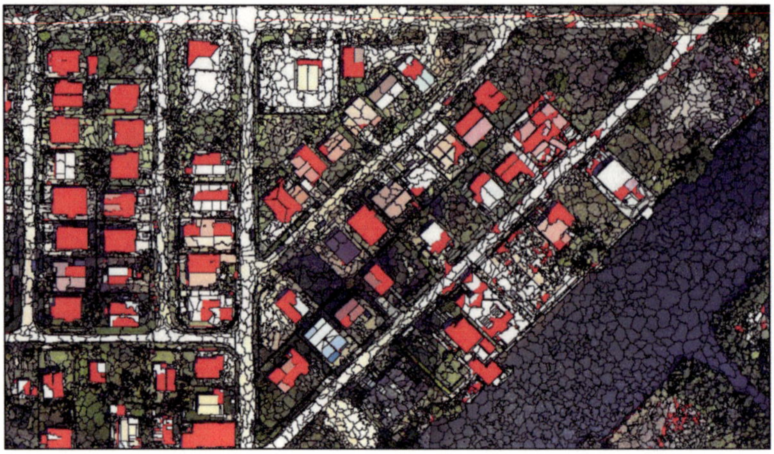

**Figure 6.**
*Building footprint extraction using PCA of RGB, NIR, slope and nDSM.*

**Figure 7.**
*Building footprint extraction using PCA of RGB, nDSM and NDVI.*

The fourth approach includes the PCA of RGB, NIR, nDSM and NDVI, a total of six bands. The NIR is introduced in the PCA to see if it can improve the extraction of buildings from other features. The result is an improvement from the third approach; however, many building outlines are still not well represented, which would require tedious editing and adjustments. Some editing tasks such as splitting polygons, and reshaping building boundaries will be exhaustive (**Figure 8**).

The fifth and final approach was conducted with the PCA of RGB, NIR and nDSM, a total of five bands. The results show a huge improvement from all other approaches in terms of selecting all features that are buildings as well as showing a well-defined boundary of building outline with a 92% extraction accuracy. Notice that there are very few other features that were classified as buildings using this approach (**Figure 9**).

**Table 2** shows a comparison of the five approaches completed. The number of all segments are the total segments of all features within the 1 sq.km of area for each approach. The segments classified as buildings are those segments that are assigned as buildings from the total of all segments. It is important to note that on average a total of six segments represents the entire outline of one building, this also depends

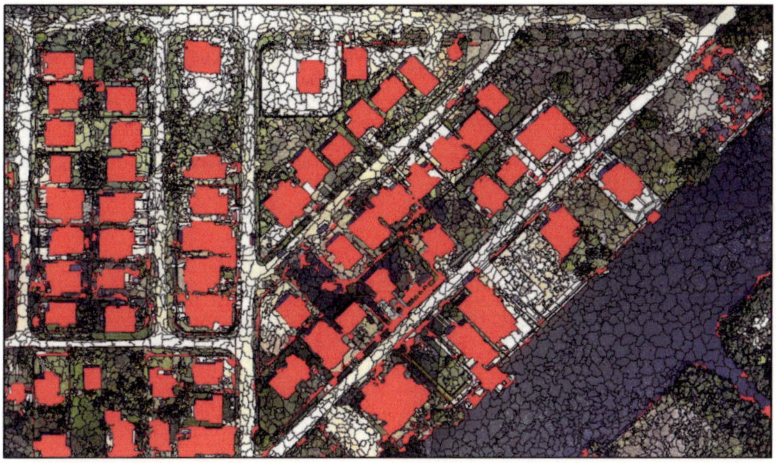

**Figure 8.**
*Building footprint extraction using PCA of RGB, NIR, nDSM and NDVI.*

**Figure 9.**
*Building footprint extraction using PCA of RGB, NIR and nDSM.*

on the size of the building. Buildings correctly classified are those buildings that are correctly classified as buildings, but their shape and outline are not properly represented. Buildings properly represented are those buildings that are correctly classified, and their shape or outline is completely represented. The percentage of the properly represented buildings was calculated from the total number (584) of actual building within the 1 sq. km area.

From all the approaches discussed above the second approach which includes slope shows to be the worst result with 28% of accuracy of extraction. The slope band does not aid in the building extraction; however, the additional band has slightly lowered the resolution of the raster data. In OBIA, the color, shape, texture, compactness, and high resolution is needed for a smooth and realistic outline of buildings. The resolution of the aerial photos is important to maintain as this was used in the segmentation process. As shown in **Figure 10**, on the left is PCA of RGB, NIR and nDSM and the image to the right is PCA of RGB, NIR, nDSM and slope. The image on the right has reduced the image resolution, this can be seen around the edges of buildings where it became fuzzy.

| PCA dataset | Number of all segments | Segments classified as buildings | Buildings correctly classified | Buildings properly represented | Percentage with actual number of buildings (584) |
|---|---|---|---|---|---|
| RGB and nDSM | 122,200 | 29,676 | 578 | 247 | 42% |
| RGB, NIR, nDSM and Slope | 189,451 | 13,796 | 443 | 163 | 28% |
| RGB, nDSM and NDVI | 164,788 | 22,946 | 584 | 326 | 56% |
| RGB, NIR, nDSM and NDVI | 18,381 | 23,332 | 584 | 357 | 61% |
| RGB, NIR and nDSM | 122,651 | 7471 | 584 | 537 | 92% |

**Table 2.**
*Comparison of the five building footprint extraction approaches.*

Segmentation at a high resolution such as 0.1 m will allow a smoother and a more defined outline of the buildings. Image segmentation completed using 1-m spatial resolution such as the LiDAR nDSM and slope will show a jagged and irregular shape of buildings.

Using a visual binary comparison method for building extraction as shown in **Table 1**, The PCA of RGB, NIR and nDSM has shown to produce the best result of all five approaches. It shows buildings are correctly classified, properly represented, and has a total of 92% accuracy of extraction from the total number of buildings within the 1 sq. km area. Looking at the table above, it is noticeable that this approach has the least number of segments assigned to buildings with 7471. The smaller number of segments allow for better classification of buildings and present very few fragments of other features that are wrongly classified as buildings.

Another evaluation of the accuracy of the extraction process was conducted using the completeness and correctness method which is also known as Area-based accuracy measures. This method measures the completeness, correctness and quality of the building extraction process. The purpose of area-based accuracy measures is to obtain stable accuracy measurements. The area-based accuracy measures (i.e., correctness, completeness, and quality) are designed for OBIA evaluation [23].

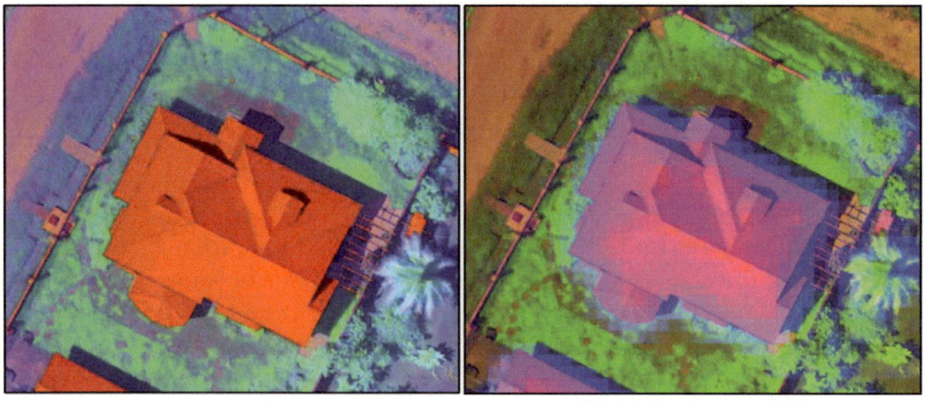

**Figure 10.**
*Comparison of PCA images resolution.*

In addition, this method can be used to calculate the commission and omission of building extraction. To complete this method, a reference building polygons and the extracted building polygons are needed. The reference data used is the 584 building polygons within the 1 sq. km area. The equation used is shown in **Figure 11**.

The completeness is the percentage of entities in the reference data that were detected, and the correctness indicates how well the detected entities match the reference data [25]. The quality of the results provides a compound performance metric that balances completeness and correctness [24]. TP (True Positive) are those areas correctly classified as buildings, FN (False Negative) are those areas that are classified as buildings but are not buildings based on the reference data. FP are those areas that are not classified as buildings during the extraction process, but are actual buildings based on the reference data. Error of commission is the same as FN, which are areas wrongly classified as buildings, and error of omission is the same as FP, which are areas that are buildings, but they are not extracted. Error of commission and omission are commonly used in the evaluation of building classification and are presented as percentages. An error of commission and omission, completeness, correctness and quality were completed for the five approaches of building extraction presented in **Table 2**. For illustration purposes the area-based accuracy measures was completed below for the PCA of RGB, NIR and nDSM using the equation in **Figure 11**. The total area for the reference data (584 buildings) is 72,360.357 sq. m. The total extracted area or classified buildings for the PCA of RGB, NIR and nDSM is 76,963.690 sq. m. The figures are illustrated below.

TP = 67454.350 sq. m. This is the correctly classified buildings in the extraction process.

FN = 9509.340 sq. m. This is the areas wrongly classified as buildings during extraction.

FP = 4906.007 sq. m. This is the areas that are buildings but are not detected as buildings.

*Completeness* = *TP/(TP + FN)* = 67454.350/(67454.350 + 9509.340) = 0.876 (**87.644%**).

*Correctness* = *TP/(TP + FP)* = 67454.350/(67454.350 + 4906.007) = 0.932 (**93.220%**).

*Quality* – *TP/(TP + FN + FP)* = 67454.350/(67454.350 + 9509.340 + 4906.007) = 0.823 (**82.392%**).

*Commission error* = *FN/TP* = 9509.340/67454.350 = 0.140 (**14.097%**).

*Omission error* = *FP/TP* = 4906.007/67454.350 = 0.072 (**7.27%**).

The calculation illustrated above was completed for the other four building extraction approaches. The result is shown in **Table 1**. Using the area-based accuracy measures, the criteria for a complete and correct building extraction are low commission and omission percentage, and high completeness, correctness, and quality percentage rate. From all five approaches, the PCA of RGB, NIR and nDSM display this criterion with low commission and omission percentage and a high percentage of completeness (87.644%), Correctness (93.220%) and high quality of

$$Completeness = TP/(TP+FN)$$

$$Correctness = TP/(TP+FP)$$

$$Quality - TP/(TP+FN+FP)$$

**Figure 11.**
*Area-based accuracy measures, source: [24].*

82.392%. In **Table 1**, it is noticeable that other approaches have higher correctness value, however their completeness and quality is poor.

There are no classification techniques that are 100% accurate, however, in OBIA, a rule-based classification can be used to improve the classification results. This was completed on the best approach discussed above involving the PCA of RGB, NIR and nDSM. The rule-based approach is only applicable where classification has been completed. It this case, it removes unwanted features that are not buildings by selecting features that have the characteristics of buildings such as size, shape, and height. Using the attribute information of classified buildings, a query is built to complete this step. The example below demonstrates this technique, where the image on the left shows features classified as buildings, looking closer at the image, the fences around these buildings are classified as buildings as well. Using the rule-based approach (**Figure 12**) this can be improved by selecting buildings within a certain height, as we know in most cases that fences are lower than houses. Therefore, a threshold is set between 6 m as the average height and 12 m as the average maximum height, this eliminates the fence as shown in the image on the right where it stays in red color, and the features that meet the criteria are selected shown in orange color.

The rule-based classification demonstrated a technique of improving the classification results by removing unwanted features based on their attribute information (**Figure 13**). However, geometrical information can be used as well, this is important where vegetations are classified as buildings. It is observed that the segments of vegetation are mostly circular in shape and the segments of buildings are rectangular. A threshold value of rectangularity can be used to eliminate vegetations that are wrong classified as buildings using their geometrical characteristics.

**Figure 12.**
*Rule-based extraction of building footprints.*

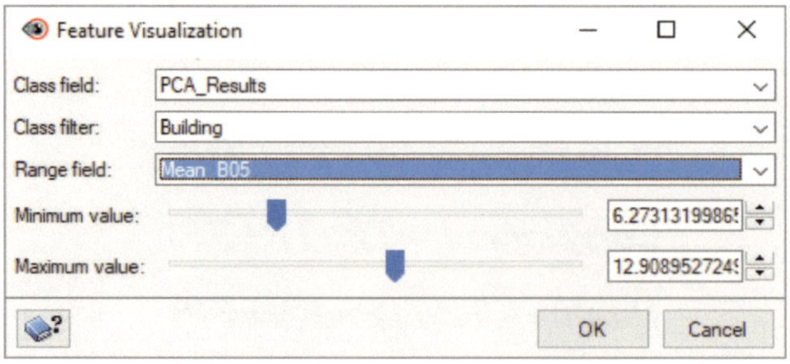


**Figure 13.**
*Building footprints extraction using attribute information.*

The result of the building extractions was converted to a feature class in ArcGIS where minimal post processing was performed. Zooming very close to a large scale of 1:250 of the building polygon you will notice that there are minor rough edges as shown in **Figure 14**. These minor rough edges cannot be seen at a scale of 1:1000.

These rough or jagged edges were eliminated using the Regularize Building Footprint tool by setting a tolerance of 0.5 m and a precision of 0.25 m, this parameter was observed to produce the best results of cleaning the edges of buildings. The result is shown in **Figure 15** where a well-defined, smooth and realistic building outline polygon is accomplished.

The building polygon was overlaid on the aerial photo and the results show a well-defined and accurate building or roof boundary (**Figure 16**).

Accurate building size and shape is important for damage assessment in flood modeling applications, as this will be used to determine the impact of a flood disaster on these structures. Ladyville village has a combination of medium and small buildings; however, it is observed that the most vulnerable populations are those that live in flood prone areas and those that live in tiny or small buildings. Proper representation of these small structures needs to be accurately represented

**Figure 14.**
*Minimal rough edges of building outlines after extraction process.*

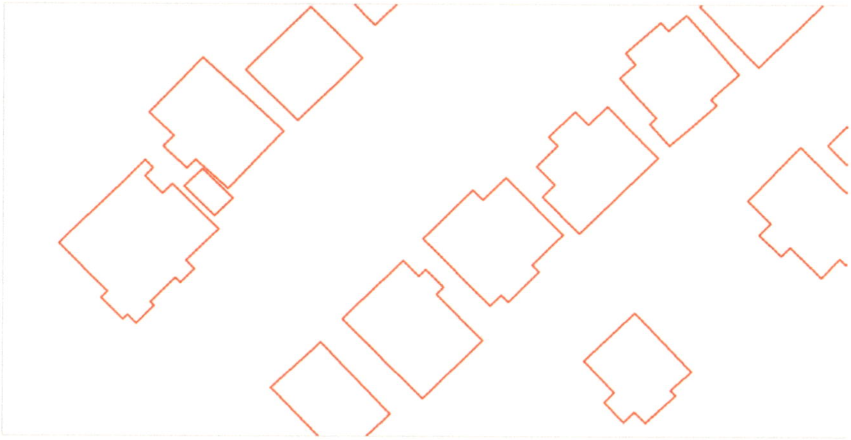

**Figure 15.**
*Regularize building footprint outlines.*

**Figure 16.**
*Building footprint outlines overlaid on aerial photos.*

for proper analysis of the extent of the damages suffered. The images of tiny houses are provided in **Figure 17**, which gives an illustration of the size of some of the buildings in the study area. What is not shown are tiny buildings that are poorly constructed and in a very dilapidated condition, which may house sometimes a family of 4 or 5 people.

During field collection and verification, it was observed that some of these small buildings were not classified in the LiDAR data. This means that their roof outline cannot be extracted. However, with OBIA process using a combination of aerial photos and Lidar height information, these small structures were successfully extracted as well. The image on the left in **Figure 18** shows small building that were not classified, with red points representing buildings. The image on the right is the result of the OBIA building extraction, which clearly shows that it has extracted

**Figure 17.**
*Example of tiny buildings not classified as buildings in LiDAR.*

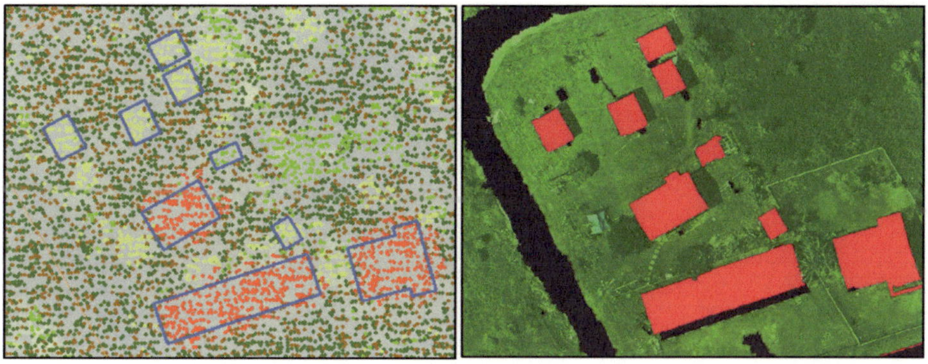

**Figure 18.**
*Tiny building footprints extraction using PCA and OBIA method.*

the roof shape of these small buildings. Small buildings in the top left illustrate this process.

This approach successfully extracts buildings from the study area, and as discussed above, minimal post processing was required. An average of 2 hours is required to complete this process. With faster computers, this time could be significantly reduced. Most of the time is spent on image preparation, PCA analysis and conversion between different raster types. This approach is a significant improvement where approximately 600 buildings can be properly represented within this period. The building's shape is well preserved. Even buildings that have a combined roof type as zinc and concrete were well outlined. The extraction process was completed at a high spatial resolution of 0.1 m (10 cm). The high resolution PCA of aerial photos and LiDAR nDSM allows the building to maintain its smooth outline with a completeness of 87.644%, Correctness of 93.220% and a quality of 82.392%.

## 5. Discussion and conclusion

A semi-automated object-based building extraction with limited post processing using the PCA image fusion technique is presented. The results show a very promising technique for precise and high-resolution extraction of buildings in urban areas using LiDAR derived height information (nDSM) combined with aerial photos (RGB and NIR). These data complement each other by providing mutual benefits in the extraction process. The RGB provided high resolution image with color which is very important in the segmentation process of OBIA to group pixels into segments, the nDSM provide height information to separate elevated structures such as buildings from other features and the NIR provides information to separate vegetation from other objects.

The extraction process was completed at a high spatial resolution of 0.1 m (10 cm). The high resolution PCA of aerial photos and LiDAR nDSM allows the building to maintain its well defined and smooth shape. The result of this study can be applied to various scenarios where accurate size and shape of buildings are important, such as in flood damage assessment.

## Acknowledgements

The author would like to acknowledge the Ministry of Works, Belize C.A for providing the LiDAR datasets and the aerial photographs for the study area.

My employer, the Ministry of Natural Resources, Belize C.A for granting me study leave to conduct this research. The Education Abroad Program at Vancouver Island University, for choosing me as a recipient of the Queen Elizabeth II Scholarship. Lastly, I would like to express my sincere gratitude to my research advisor Dr. Michael Govorov. Thank you for your guidance and support throughout this project.

## Author details

Alfred Cal
Vancouver Island University, Nanaimo, BC, Canada

*Address all correspondence to: cal_alfred@yahoo.com

## References

[1] Du S, Zhang Y, Zou Z, Xu S, He X, Chen S. Automatic building extraction from LiDAR data fusion of point and grid-based features. ISPRS Journal of Photogrammetry and Remote Sensing. 2017;**130**:294-307. DOI: 10.1016/j.isprsjprs.2017.06.005

[2] Gharbia R, Azar AT, El Baz A, Hassanien AE. Image fusion techniques in remote sensing. 2014. [Online]. Available from: http://arxiv.org/abs/1403.5473

[3] Zeng C. Automated building information extraction and evaluation from high-resolution remotely sensed data. Electronic Thesis and Dissertation Repository. 2014:2076. Available from: https://ir.lib.uwo.ca/2076

[4] Weih RC, Riggan ND. Object-based classification vs. pixel-based classification: Comparitive importance of multi-resolution imagery. International Archives of the Photogrammetry, Remote Sensing and Spatial Information Sciences - ISPRS Archives. 2010:38

[5] Blaschke T. Object based image analysis for remote sensing. ISPRS Journal of Photogrammetry and Remote Sensing. 2010;**65**(1):2-16. DOI: 10.1016/j.isprsjprs.2009.06.004

[6] Awrangjeb M, Fraser CS. Automatic segmentation of raw LIDAR data for extraction of building roofs. Remote Sensing. 2014;**6**(5):3716-3751. DOI: 10.3390/rs6053716

[7] Ren Z, Zhou G, Cen M, Zhang T, Zhang Q. A novel method for extracting building from LIDAR data-Fc-S method. International Archives of the Photogrammetry, Remote Sensing and Spatial Information Sciences: ISPRS Archives. 2008;**37**:283-288

[8] Wang Z, Schenk T. Extracting building information from LIDAR data. International Archives of the Photogrammetry, Remote Sensing. 1998;**32**(3/1):279-284. Available from: http://citeseerx.ist.psu.edu/viewdoc/download?doi=10.1.1.50.8154&rep=rep1&type=pdf

[9] Priestnall G, Jaafar J, Duncan A. Extracting urban features from LiDAR digital surface models. Computers, Environment and Urban Systems. 2000;**24**(2):65-78. DOI: 10.1016/S0198-9715(99)00047-2

[10] Yong LI, Huayi WU. Adaptive building edge detection by combining Lidar data and aerial images. International Archives of the Photogrammetry, Remote Sensing and Spatial Information Sciences. 2007;**XXXVII**(B1):197-202

[11] Tiwari PS, Pande H, Nanda BN. Building footprint extraction from ikonos imagery based on multi-scale object oriented fuzzy classification for urban disaster management. International Archives of the Photogrammetry, Remote Sensing and Spatial Information Sciences. 2006;**34**:1-7

[12] Feng T, Zhao J. Review and comparison: Building extraction methods using high-resolution images. In: 2nd International Symposium on Information Science and Engineering, ISISE 2009. 2009. pp. 419-422. DOI: 10.1109/ISISE.2009.109

[13] Attarzadeh R, Momeni M. Object-based building extraction from high resolution satellite imagery. International Archives of the Photogrammetry, Remote Sensing and Spatial Information Sciences, ISPRS. 2012;**XXXIX-B4**(September):57-60. DOI: 10.5194/isprsarchives-xxxix-b4-57-2012

[14] Rottensteiner F, Summer G, Trinder J, Clode S, Kubik KKT.

Evaluation of a method for fusing LIDAR data and multispectral images for building detection. In: Joint workshop of ISPRS and DAGM-CMRT05. Vol. 36. ISPRS. 2005. No. Part 3/W24, pp. 15-20. Available from: papers2://publication/uuid/D641830A-1713-4ADE-949A-7C008A05F110

[15] Awrangjeb M, Ravanbakhsh M, Fraser CS. Automatic detection of residential buildings using LIDAR data and multispectral imagery. ISPRS Journal of Photogrammetry and Remote Sensing. 2010;**65**(5):457-467. DOI: 10.1016/j.isprsjprs.2010.06.001

[16] Gerke M, Xiao J. Fusion of airborne laserscanning point clouds and images for supervised and unsupervised scene classification. ISPRS Journal of Photogrammetry and Remote Sensing. 2014;**87**(March):78-92. DOI: 10.1016/j.isprsjprs.2013.10.011

[17] Moeller MS, Blaschke T, Urban change extraction from high resolution satellite image. ISPRS Technical Commission. II Symposium Vienna. 2006

[18] F. Analysis and S. Econometrics. Spatial autoregressive analysis; spatial econometrics

[19] Adam HE, Csaplovics E, Elhaja ME. A comparison of pixel-based and object-based approaches for land use land cover classification in semi-arid areas, Sudan. IOP Conference Series: Earth and Environmental Science. 2016;**37**(1):0-10. DOI: 10.1088/1755-1315/37/1/012061.

[20] Syed S, Dare P, Jones S. Automatic classification of land cover features with high resolution imagery and Lidar data: An object-oriented approach. Image (Rochester, N.Y.). 2005:512-522

[21] Khatriker S, Kumar M. Building footprint extraction from high resolution satellite imagery using segmentation. International Archives of the Photogrammetry, Remote Sensing and Spatial Information Sciences: ISPRS Archives. 2018;**42**(5):123-128. DOI: 10.5194/isprs-archives-XLII-5-123-2018

[22] Blaschke T, Lang S, Hay GJ. Object-based image analysis. Object-based Image Analysis Spatial Concepts for Knowledge-Driven Remote Sensing Applications. January 2008;**2008**: V–VIII. DOI: 10.1007/978-3-540-77058-9

[23] Cai L, Shi W, Miao Z, Hao M. Accuracy assessment measures for object extraction from remote sensing images. Remote Sensing. 2018;**10**(2). DOI: 10.3390/rs10020303

[24] Rutzinger M, Rutzinger M, RottensteinerF, RottensteinerF, PfeiferN. A comparison of evaluation techniques for building extraction from airborne laser scanning. IEEE Journal of Selected Topics in Applied Earth Observations and Remote Sensing. 2009;**2**(1):11-20. DOI: 10.1109/JSTARS.2009.2012488

[25] Uzar M. Automatic building extraction with multi-sensor data using rule-based classification. European Journal of Remote Sensing. 2014;**47**(1):1-18. DOI: 10.5721/EuJRS20144701